Ion and
Atomic Beams for
Controlled Fusion
and Technology

Ion and Atomic Beams for Controlled Fusion and Technology

M. D. Gabovich
Ukrainian Institute of Physics
Kiev, USSR

N. V. Pleshivtsev *and*
N. N. Semashko
Kurchatov Institute
Moscow, USSR

Translated from Russian by

Donald H. McNeill

CONSULTANTS BUREAU
NEW YORK AND LONDON

Library of Congress Cataloging in Publication Data

Gabovich, Mark Davidovich.
[Puchki ionov i atomov dlía upravlíaemogo termoíadernogo sinteza i tekhnologicheskikh tseleĭ. English]
Ion and atomic beams for controlled fusion and technology / M. D. Gabovich, N. V. Pleshivtsev, and N. N. Semashko; translated from Russian by Donald H. McNeill.
p.   cm.
Translation of: Puchki ionov i atomov dlía upravlíaemogo termoíadernogo sinteza i tekhnologicheskikh tseleĭ.
Includes bibliographical references and index.
ISBN 0-306-11016-4
1. Ion bombardment. 2. Atomic beams. I. Pleshivtsev, N. V. (Nikolaĭ Vasil'evich) II. Semashko, N. N. (Nikolaĭ Nikolaevich) III. Title.
QC702.7.B65G3313   1988   88-25607
537.5′6—dc19   CIP

This translation is published under an agreement with the Copyright Agency of the USSR (VAAP).

© 1989 Consultants Bureau, New York
A Division of Plenum Publishing Corporation
233 Spring Street, New York, N.Y. 10013

Printed in the United States of America

PREFACE

A beam of ions in the form of "canal rays" was first observed in 1886 by E. Goldstein. The first ion source was invented by J. J. Thomson in 1910. This ion source became the basis for the first widespread application of ion sources in mass spectrographs and mass spectrometers. The second important application of ion sources is ion accelerators, which since the beginning of the 1930s have been employed in research on nuclear reactions and are now used in industry and medicine.

A third application of ion sources is in systems for isotope separation and research on the interaction of atomic particles with solids (1940s). The result of this research and development is the use of ion sources in semiconductor doping, decontamination of surfaces, and micromachining of surfaces (1960s and 1970s), which is a fourth area of applications for ion sources.

The heating of plasmas in magnetic confinement devices to thermonuclear temperatures (100-1000 MK) with the aid of megawatt beams of hydrogen and deuterium ions and atoms has become a fifth promising area of application for ion sources which can produce ion beams with steady-state currents of up to 100 A.

Finally, experimental and industrial research are under way on the alloying of metals and the fabrication of coatings which greatly improve the physical and chemical properties of metals. These coatings can increase the hardness, high-temperature corrosion resistance, and wear resistance of metals, and can enhance or reduce friction, etc.

Since ion sources are now used in many areas of science, engineering, technology, and medicine and new possibilities are being opened up for them, the physical and technical fundamentals of modern ion sources and their current and future technological applications will be of some interest to readers.

We do not examine ion implantation in metals here. There is a very extensive literature on this topic, primarily in the form of scientific articles and the proceedings of conferences and meetings.

Several thousand publications already exist on the subjects touched upon in this book. For this and other reasons we have been forced in many instances to make indirect references to reviews, monographs, and conference proceedings by mentioning the authors' names in the text.

Fluxes and beams of different elements are gaining wider use in engineering and technology. This is because of the need to raise the accuracy with which parts and components are fabricated, to improve their physical, chemical, and service properties, to reduce the expenditure of hard-to-obtain and expensive materials, to fully automate engineering processes, and, finally, because of the miniaturization of components for applications in outer space and daily life. Ion sources, sputtering processes, and ion implantation can be used to solve these problems.

Chapters 1, 2, and 9 were written by M. D. Gabovich, Chapters 4-8 by N. V. Pleshivtsev, and Chapter 3 by N. N. Semashko. The authors thank their colleagues and the staff at the ONTI (United Scientific and Technical Presses) and the library of the Kurchatov Institute of Atomic Energy for providing information, the staff at the Division of Plasma Physics of the BNTI (Bureau of Scientific and Technical Information) for help in the layout of the manuscript, and the official reviewer, Dr. V. E. Minaichev, for valuable comments.

The Authors

CONTENTS

Chapter 1

ION SOURCES, PLASMA EMITTERS, AND THE PRIMARY FORMATION OF ION BEAMS

1.1 Plasma Ion Sources and Their Major Parameters

An ion source is a device for producing an ion beam, i.e., a shaped flux of ions in which the velocity of directed motion is significantly greater than the random thermal velocity of the particles. Here we describe gaseous discharge devices for generating ions. In these sources, ions are collected from the surface of a gaseous discharge plasma by an external electric field. We shall also examine the properties of plasma emitters independently of the means by which they are created, as well as the shaping of ion beams extracted from them. The physical principles involved in obtaining the required highly ionized, dense plasma with a large emitting surface are described in Chapter 3 and are discussed in detail in the books by Gabovich [1] and Semashko et al. [2].

Applications of ion sources. Plasma ion sources are now widely used in science and technology, for example, in

1) charged-particle accelerators for nuclear physics research and the production of intense neutron fluxes;

2) quasistationary injectors of fast neutral atoms (neutral beam injectors) for fueling and heating of magnetically confined thermonuclear plasmas;

3) pulsed systems for inertial confinement fusion;

4) ion engines;

5) devices for electromagnetic separation of isotopes;

6) research on the physics of atomic collisions and the interaction of fast particles with solids, plasma diagnostics, and the elemental and isotopic analysis of solid surfaces;

7) various technological processes, e.g., precision machining of solids, doping of semiconductors and metals to change their physical and chemical properties, welding, passivation of metal surfaces, removal of contamination from or deposition of coatings on metal surfaces by means of cathode sputtering, production of thin films by deposition of ions or fast atoms, etc.; and

8) pumping of lasers.

Naturally, each of these applications places specific demands on an ion source. For example, the ion sources used in electrostatic accelerators are not meant to yield high-current ion beams, but must have a long operating lifetime. This last point is important because in these accelerators the ion source is usually located near the high-voltage electrode in a compressed gas, and replacement or repair of the source requires that the generator be opened up, an action that is highly undesirable. In tandem accelerators, on the other hand, the ion source is not kept at a high potential, and so energy supply to, as well as cooling, disassembly, and repair of, the source are easier. These accelerators typically have a negative ion source, the accelerated ions from which lose two electrons in a charge exchange cell and undergo further acceleration as positive ions.

In a cyclotron the ion source is usually located within a limited space within a strong magnetic field and the ions are extracted perpendicular to the magnetic field. From this it follows, in particular, that the source design must include forced cooling of a quite durable cathode, since the latter may be deformed owing to the interaction of the heater current with the external magnetic field. When ions are injected into a cyclotron from an external source along the magnetic field with a subsequent 90° change in their trajectory, many of the limitations usually imposed on an ion source located in the interpolar gap no longer apply. Multiply charged ion sources are extensively used in cyclotrons, and negative ion sources are sometimes used. The development of intense sources of multiply charged ions is an important problem.

Sources for electromagnetic separation of isotopes of refractory elements differ greatly from other sources in that the gaseous discharge medium is not an easily introduced gas but the vapor of these elements, which sometimes must be obtained by keeping the entire source at a temperature above 2000°C. Sources of refractory element ions are also needed for various accelerators and technological purposes.

Quasistationary neutral beam injectors now typically have positive ion sources with currents on the order of a hundred amperes or negative ion sources with currents on the order of ten amperes. Certain difficulties arise in the production of beams with such high currents. These difficulties originate in the need for careful development of the systems for primary shaping and acceleration, as well as of the entire injector channel. Complicated ion-optics problems must be solved, the heat released in the discharge volume or on the surfaces of various electrodes must be dissipated, and cathode sputtering of the electrodes must be prevented. Transport difficulties are common in the pulsed ion beams with currents of hundreds of kilo-amperes employed in inertial confinement fusion systems.

A range of ion currents (encompassing ten orders of magnitude), different ion species (positive, negative, and multiply charged ions with charges up to those of "bare" nuclei, ions of any element in the periodic table, atomic and molecular ions, charged clusters), and the need to make parameters match the required purpose of the apparatus are factors which determine the variety of existing plasma ion sources and their complexity, which is often hidden by an apparent simplicity. Some basic requirements for these devices can be formulated only in a general form:

1) an ion source with a structure for primary acceleration of ions and forma-tion of a beam must produce a stationary or pulsed ion beam with the required cur-rent and energy and ion-optical parameters which will allow it to be used in the best way;

2) the ion source must yield an ion beam with a definite composition, i.e., must deliver ions with a given mass and charge; other components in the beam are undesirable since they "contaminate" the beam, increase the load on the power sup-plies, reduce the electrical resistance of the accelerator tube, etc.;

3) the ion source must operate stably since modulation of the ion beam current is generally undesirable;

4) a given beam must be obtained with a minimal expenditure of working material because a high vacuum must be maintained in the region of primary beam production and acceleration in order to enhance its electrical breakdown strength; when expensive gases or metals are used, this requirement also follows from eco-nomic considerations;

5) the power expended in supplying the source should be minimal;

6) the source must be sufficiently reliable in operation and as simple as possi-ble in design, supply, and control;

7) the source must have a sufficiently long continuous operating period and as long an operating lifetime as possible; and

8) the beam extracted from the source must have a small spread in ion ener-gies and permit the required focusing and transport.

The parameters of ion sources. In accordance with the requirements listed above, an ion source is characterized by the following most important parameters:

1) the total focusable ion current I_+ and the corresponding current density $j_+ = I_+/S_0$, where S_0 is the cross-sectional area of the output aperture of the source. The energy of the beam ions E is numerically equal to the accelerating voltage U times the ionic charge Ze;

2) the efficiency, or ion current, per unit power W_{source} delivered to the source is $H = I_+/W_{source}$, mA/W. This power includes the discharge power and the power expended in heating the cathode, in maintaining a magnetic field, and so on. Sometimes the concept of energy efficiency, defined as the ratio of the beam power to the power from all the power supplies, $\xi = I_+U/W_{source}$, is used;

3) the beam perveance $P = I_+/U^{3/2}$;

4) the gas efficiency, given by the ratio of the number of atoms converted into beam ions to the number of atoms fed into the source;

5) the relative proton content in a hydrogen ion beam (for proton sources), $\beta = I_{H_1}^+/\Sigma_{i=1,2,3}I_{H_i}^+$;

6) the degree of modulation of the beam $M = \Delta I_+/I_+$, where ΔI_+ is the fluctuating component of the ion beam current; and

7) the divergence angle of the shaped beam and the emittance and brightness of the source. The concept of emittance warrants special attention [3].

If special measures are not taken, a beam will spread out as it propagates. One of the reasons for this is the action of the space charge of the beam (see Chapter 2). Another reason for spreading or for difficulties in focusing is the random scatter in the transverse thermal velocities of the ions. The concept of phase space is used to characterize the effect of this scatter. In general, the state of a single particle at every moment is characterized by a point that represents this particle in a $2n$-dimensional space, where n is the number of degrees of freedom. Of the $2n$ variables, half characterize the velocity of the particle for each degree of freedom (the momenta p_n) while the remaining n variables characterize its position in space (the coordinates q_n). The values of the generalized coordinates at given times are plotted in phase space, and the path of a point in phase space (the phase trajectory of a particle) describes the change in a particle's state with time. Representative points in the beam occupy a particular volume in phase space known as the phase volume of the beam. Because of the scatter in velocity and position of the particles, this volume is finite. According to Liouville's theorem, which assumes, among other things, that there is no dissipation, the six-dimensional phase volume is invariant. In general, a projection of this volume onto the y, p_y plane or other planes may not be invariant. A change in the two-dimensional projection of the phase volume onto one of the coordinate planes causes a change in the corresponding projections on other planes. In particular cases the two-dimensional transverse phase volume, given by

$$V_{ph} = \frac{1}{\pi m_0 c} \int dy\, dp_y, \qquad (1.1)$$

may be conserved (the transverse and longitudinal motions are not coupled). The concept of emittance is often used to characterize the phase volume of a beam. The emittance

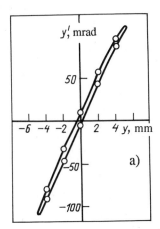 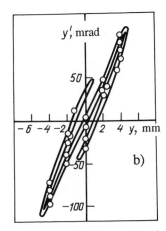

Fig. 1.1. Emittance diagrams for an rf source with (a) an extraction voltage of 15 kV, beam current of 90 mA, and $\mathscr{E} = 37$ mm·mrad, and (b) an extraction voltage of 20 kV, beam current of 150 mA, and $\mathscr{E} = 69$ mm·mrad.

$$\mathscr{E} = \frac{1}{\pi} \int dy \, dy' \qquad (1.2)$$

represents the area occupied by representative points of the beam on the y, $y' = dy/dz = p_y/p_z$ plane divided by π; $\mathscr{E}$ and V_{ph} are related by $\mathscr{E} = V_{ph}/(\beta\gamma)$, where $\gamma = 1/\sqrt{1 - \beta^2}$ and $\beta = v/c$. The directly measured quantity is the emittance. One feature of the two-dimensional phase volume is that it is independent of the energy of the particles and it is sometimes called the normalized emittance. (The emittance decreases when the energy of the beam particles is raised.) Under certain conditions, the reduced emittance and the source plasma parameters are related by

$$V_{ph} = 2r_0 \left(2 \chi \, T_i / (Mc^2)\right)^{1/2}, \qquad (1.3)$$

where r_0 is the radius of the plasma which serves as a source of ions with temperature T_i.

Several examples can be introduced to illustrate the concept of emittance. If all the particles in a beam move strictly parallel to the z axis, then the representative points lie on the abscissa of the coordinate system y, y' and occupy a segment of length equal to the beam diameter. In this case the phase volume of the beam is zero. The phase volume of a homocentric beam is also equal to zero. If, on the other hand, the transverse velocities of the beam particles at given points are distributed over certain finite intervals, then the phase volume is finite. In the y, y' plane the phase volume may be bounded, for example, by an ellipse. A converging beam corresponds to an ellipse with a negative slope of the major axis and a diverging beam to an ellipse with a positive slope of the same axis. As an example, Fig. 1.1 shows experimentally obtained phase "portraits" of beams from an rf ion source under different operating conditions. Both the dimensions and the shape of the plasma boundary affect the emittance. The characteristic S-shaped projection of the phase volume (see Fig. 1.1b) is related to a complicated plasma boundary shape. The emittance may increase as a result of periodic changes in the shape and location of the plasma boundary owing to different kinds of oscillations in the source plasma.

In order to obtain well-focused ion beams with a high current density, the ion source must have a high brightness, which is usually defined as the ion current density per unit solid angle. An analog of this quantity is the brightness given by the ratio of the beam current to the four-dimensional phase volume. For an axially symmetric beam the brightness can be written in the form

$$B = \frac{I_+}{(\pi^2/2)\, V_{ph}^2} = \frac{I_+}{4\pi^2 r_0^2}\left(\frac{Mc^2}{T_i}\right). \tag{1.4}$$

1.2. Plasma Emitters of Positive Ions

The plasma in a real ion source is not an isotropic medium since, in particular, directed flows of charged particles exist in it and it is acted on by an external magnetic field. However, for simplicity, if neutral atoms are neglected, a plasma emitter is often regarded as a uniform mixture of two gases, the ions and electrons, with roughly equal densities $n_+ \approx n_e$ but different temperatures T_+ and T_e. The large mass difference between the ions and electrons makes energy exchange between them difficult. Therefore, although each gas may be in thermal equilibrium by itself, the electron and ion gases are not usually in equilibrium in the cases of interest to us and $T_e > T_+$.

Electron extraction from the surface of a thermionic emitter. In order to better understand the features of ion extraction from a plasma emitter, we may compare this process with the emission of electrons from a heated cathode. When the potential difference U_0 between two parallel electrodes, one of which is a thermionic cathode, is raised, the potential distribution and current change. Under the influence of the intrinsic space charge of the electrons, a potential profile with a characteristic minimum is first formed (Fig. 1.2a). The emitted electrons are not monoenergetic, but have an energy distribution with a temperature equal to the cathode temperature T_K. Only those electrons whose energy is greater than the potential barrier $e\Delta U_{min}$ reach the anode, while the others return to the cathode. This is the mechanism by which the intrinsic space charge limits the electron current. The dependence of the current density on the anode potential in this regime is found by solving Poisson's equation

$$\partial^2 U/\partial z^2 = -4\pi\rho_e \tag{1.5}$$

with the appropriate boundary conditions. In particular, at $z = z_{min}$, $\partial U/\partial z = 0$. However, for an approximate solution of Eq. (1.5) it is possible to use the conditions $U = 0$ and $\partial U/\partial z = 0$ at $z = 0$ and $U = U_0$ at $z = d$. The resulting errors are evidently small if U_0 is sufficiently large, i.e., $\Delta U_{min} \ll U_0$ and $z_{min} \ll d$.

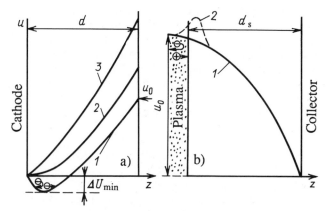

Fig. 1.2. The potential distribution in a diode with a heated cathode for different values of the anode potential (a, curves 1–3) and in the layer between a plasma and a negatively biased collector (b, curve 1; a distribution with a maximum, i.e., curve 2, cannot exist).

If we neglect the thermal velocities of the electrons, then the electron space charge density ρ_e and j_e are related by

$$j_e = -\rho_e \sqrt{2eU/m} \,. \tag{1.6}$$

Substituting ρ_e in Eq. (1.5), we obtain

$$\frac{\partial}{\partial z}\left(\frac{\partial U}{\partial z}\right)^2 = \frac{8\pi j_e}{\sqrt{2e/m}} \frac{1}{\sqrt{U}} \frac{\partial U}{\partial z} \,. \tag{1.7}$$

The first integral of Eq. (1.7) yields

$$\partial U/\partial z = \sqrt{16\pi j_e / \sqrt{2e/m}} \; U^{1/4}. \tag{1.8}$$

The second integral yields an expression for the potential distribution:

$$U = (9\pi/\sqrt{2e/m})^{2/3} j_e^{2/3} z^{4/3}. \tag{1.9}$$

Substituting the second of the above boundary conditions, we obtain the well-known "three-halves" (Child–Langmuir) law, the desired expression for the space charge current density limit:

$$j_e = \frac{1}{9\pi} \sqrt{\frac{2e}{m}} \frac{U_0^{3/2}}{d^2} \,. \tag{1.10}$$

The growth in the current density as the potential is raised is related to the accompanying reduction in the depth of the potential barrier ΔU_{min}.

For a certain value of U_0 the field goes to zero right at the cathode (curve 2 of Fig. 1.2a). Under these conditions, and when U_0 is increased further, there is no field to slow down the electrons over the entire interelectrode gap, so that all the emitted electrons reach the anode and the current in the circuit equals the cathode emission current. A transition from a space-charge-limited current to a saturation current takes place.

Ion extraction from a plasma emitter. Positive ions are extracted in a different manner from a plasma emitter, which is a source of charges of both signs. Suppose a collector with a negative potential relative to the plasma is placed in the plasma. When the potential is large enough, the electrons return to the plasma and the collector collects only positive ions. A layer of moving ions develops between the collector and the resulting plasma boundary (Fig. 1.2b). By definition, the plasma is quasineutral ($n_+ \approx n_e$) and the electric field in it is very small ($dU/dz \approx 0$). Outside the plasma there is an ion space charge moving toward the collector. However, in this case a potential maximum analogous to the minimum for the electrons, which would serve as a potential barrier for the ions and limit their current, cannot develop. It would quickly be flattened out by an influx of electrons from the plasma emitter.

When ions are extracted from a plasma, therefore, there is no intrinsic space charge limit to the current. All the ions moving from the plasma to its boundary reach the collector; i.e., a saturation current always flows. The current density (A/cm^2) of ions escaping the plasma, or the "ion emission current of the plasma," is proportional to the ion density and to the velocity they acquire at the plasma boundary:

$$j_+ = 0.4 e n_+ (2\chi T_e/M)^{1/2} = 8 \cdot 10^{-16} n_+ \left(\frac{T_e}{\mu}\right)^{1/2}. \tag{1.11}$$

The ion emission depends on the electron rather than the ion temperature because near the boundary of a plasma there is an electric field which returns electrons to the plasma and, therefore, accelerates ions in this region. Hence, the ions acquire an initial velocity which depends on T_e. Despite the absence of space-charge current limits, the "three-halves" (Child–Langmuir) law is applicable to ion collection from a plasma emitter since the electric field is zero at its boundary. As in Eq. (1.10), this law can be written in the form

$$j_+ = \frac{1}{9\pi}\left(\frac{2e}{M}\right)\frac{U_0^{3/2}}{d_s^2}, \tag{1.12}$$

where M is the ion mass and d_s is the width of the ion space-charge layer (sheath) separating the ion collector from the plasma boundary. Now, however, a change in U_0 has no effect on the ion current density (which according to Eq. (1.11) is a function of the plasma parameters alone) but only affects d_s, i.e., causes a displacement in the plasma boundary:

TABLE 1.1. Ion Densities and Minimum Gas Pressures Corresponding to Given Values of the Ion Current Density

n_+, cm^{-3}	10^{11}	10^{12}	10^{13}	10^{14}	10^{15}
j_+, A/cm^2	0.025	0.25	2.5	25	250
p_{min}, Pa	$4 \cdot 10^{-4}$	$4 \cdot 10^{-3}$	$4 \cdot 10^{-2}$	$4 \cdot 10^{-1}$	4

$$d_s = \frac{1}{(9\pi)^{1/2}} \left(\frac{2e}{M}\right)^{1/4} \frac{U_0^{3/4}}{j_+^{1/2}}. \tag{1.13}$$

As an example, when a negative potential on the collector is raised, there is a rise in the positive ion charge in the sheath volume which shields the collector, i.e., the thickness of the sheath increases.

Shape and location of the plasma boundary. The main feature of a plasma ion emitter, therefore, is the absence of a space-charge current limit [1]. The location and, under actual conditions, the shape of a plasma emitter, as well, depend strongly on the ion emission current density j_+ and potential difference U_0 localized in the ion space-charge layer. It should be noted that because of the various instabilities in which a plasma is so rich, the ion current density and, consequently, the shape and location of a plasma emitter may vary with time and thereby have an additional unfavorable effect on the qualitative characteristics (i.e., the emittance) of the extracted beam.

Table 1.1 shows the ion current densities given by Eq. (1.15) for fixed ion densities with $\mu = 1$ and $T_e = 10^5$ K. The table also lists the minimum gas pressure necessary to obtain the specified ion density under conditions of full ionization. This table shows, for example, that in continuous ion sources with a current density $j_+ = 10\text{-}100$ A/cm^2 it is necessary to maintain a plasma with an ion density of 10^{14}-10^{15} cm^{-3}. Thus, high current-density ion sources require efficient methods for creating plasmas with high ion densities. This, however, is not sufficient for making an efficient ion source.

Enhancing the ion utilization efficiency. Consider a spherical (for simplicity) ion source vessel of radius R containing a uniform plasma with an ion current density j_+ to the wall from which ions are extracted through an aperture of area $S_0 = \pi r_0^2$ so that the plasma boundary remains close to a spherical surface. The fraction of ions that is used is then $\alpha = j_+ S_0/(j_+ S) = r_0^2/(4R^2)$. For the typical $r_0 = 10^{-1}$ cm and $R = 1\text{-}10$ cm, this fraction is very small: $\alpha = 10^{-3}$-10^{-5}. The efficiency of utilization of the ions produced in an ion source can be raised by: (1) increasing the area of the emitting surface $S_{em} \gg S_0$ by introducing a field in the source vessel which removes ions; (2) creating a nonuniform plasma such that the most intense plasma is concentrated near the outlet aperture; and (3) making the values of r_0 and R closer. The first method is used in rf sources with low ion densities. The second is realized, for example, in a duoplasmatron by narrowing the discharge region and further contracting the discharge with a magnetic field. The third method is typical of the sources without a magnetic field that are used in fast atom injectors. There the area S_0 is large compared to the surface of the vessel walls which remove ions unproductively.

It is important to note that under certain conditions the ion current density collected from the surface of a plasma emitter may be greater than that given in Eq. (1.11). If the ion emitter is a plasma formed in a plasma accelerator [4] and propagates with a velocity v_{pl} greater than the ion-acoustic speed $v_s \approx (\chi T_e/M)^{1/2}$, then the ion current density is given by

$$j_+ = e n_+ v_{pl}.$$

(1.14)

There is yet another way to increase the current density extracted from the surface of a plasma emitter [5]. Using the principle of plasma optics that the magnetic field lines in a plasma are equipotentials, it is possible to artificially create a flux of ions to the extraction target by creating an internal electric field in the plasma. This possibility, which is of fundamental importance, has been demonstrated using PIG discharge. If the source anode consists of two parts A_1 and A_2, then for a suitable potential distribution in the cathode layer, the discharge column will also consist of two parts with potentials close to the potentials of the anodes A_1 and A_2. Thus, for an appropriate potential difference ΔU between the anodes, an electric field $E_\perp$ perpendicular to the magnetic field will appear in the plasma and accelerate ions toward the slit. A potential difference $\Delta U \approx 20$ V is enough to raise the extracted ion current by a factor of 3-4. Changing the sign of ΔU will, as is to be expected, reduce the extracted ion flux.

An efficient source is distinguished both by a high efficiency of utilization of the ions that are produced, α, and by a high gas efficiency, enhancements in which help to raise the degree of ionization of the gas in the source.

The characteristics of negative ion extraction. During extraction of positive ions from plasma emitters the other component (the electrons) is returned to the plasma by the external electric field. When negative ions are extracted, however, the ion flux is always accompanied by a considerably more intense flux of charges of the same sign, i.e., electrons. The accompanying electrons load the power supplies, cause unacceptable overheating of the electrodes, and disrupt normal operation of the ion beam formation system. Thus, every plasma source of negative ions has an electron filter.

1.3   Formation of Ion Beams.   Plasma Focusing

One of the most important problems in developing an ion source is making the correct design choice for the system for forming the ion beam. In the case of gaseous discharge sources of charged particles, this problem is considerably more complicated than forming a beam emitted from a solid surface. Unlike a thermionic emitter, a gaseous discharge is an intense source of neutral gas and the energy of the particles leaving the plasma boundary is usually greater than that of the particles

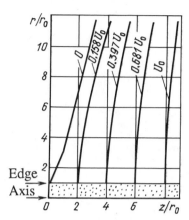

Fig. 1.3. A family of equipotential surfaces in a gun that forms a parallel axially symmetric beam.

emitted by a thermionic source. And after all, the fundamental feature of ion or electron beam formation from a gaseous discharge is that one of the electrodes, the immediate source of charged particles, is not a fixed surface (such as a thermionic cathode or solid surface on which ions are formed) but is the boundary of a plasma, whose location and shape depend on the plasma parameters and on external fields. These systems are known as plasma focusing systems, and they may be classified as follows:

1. Systems in which the plasma boundary (the plasma emitter) is inside the gaseous discharge vessel of the source. This method of extracting ions and forming a beam is mostly used in rf ion sources and is suitable for producing ion beams of relatively low intensity. The maximum beam current passing through a channel of radius r_0 and length L in an extraction electrode is given by [see Eq. (2.15)]

$$I_{max} = 1.08 \cdot 2^{3/2} (e/M)^{1/2} U_0^{3/2} (r_0/L)^2, \qquad (1.15)$$

where U_0 is the negative potential of the extraction electrode. Thus, for a given angular divergence of the beam at the channel outlet $[r_0/L = (1/2)(dr/dz)|_{out}]$ and a fixed particle energy, the maximum current is independent of the dimensions of the channel. This limit on the current of the shaped beam can be avoided if n channels are made instead of one. Then the overall maximum current has been shown by Serbinov and Morok [1] to be $I_+ = nI_{max}$.

2. In a whole series of widely used low-emittance sources the plasma boundary lies in the plane of a small output aperture or near this plane [14]. For a certain shape of the electrode formed by the wall with the output aperture and of the extractor electrode which removes the ions, the boundary conditions for a so-called Pierce system may be satisfied. The equipotentials and, therefore, the shape of the electrodes for an ion gun that forms a rectilinearly propagating cylindrical beam are shown in Fig. 1.3. An important feature of the electric field in the gun is that at any

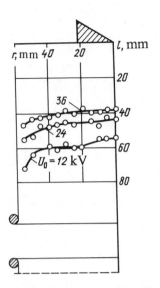

Fig. 1.4. The shape and location of the boundary of the penetrating plasma for different potentials on an immersion lens and for a single central emission aperture.

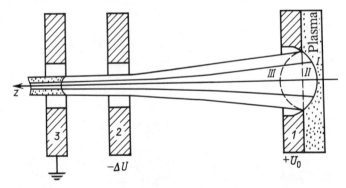

Fig. 1.5. A system for forming an ion beam with plasma focusing: III, II, and I denote the locations of the plasma boundary as U_0 is raised.

point on the boundary between the beam and the charge-free region, the potential gradient $\partial U/\partial r = 0$, which also makes the beam propagate rectilinearly. For all Pierce guns that form parallel or converging beams the angle between the zero equipotential and the normal to the emitter surface is $67.5°$. The potential distribution in the beam formed by the gun shown in Fig. 1.3, as in a one-dimensional flow of unbounded cross-sectional area [see Eq. (1.9)], has the form

$$U = (9\pi)^{2/3} (2e/M)^{-2/3} j_+^{2/3} z^{4/3}. \tag{1.16}$$

The electric field strength near the extractor $(z = d)$ is

$$E_d = 4U_0/(3d) = 2^{7/4} \pi^{1/2} (M/e)^{1/4} j_+^{1/2} U_0^{1/4}. \tag{1.17}$$

The current density and total current are given by

$$j_+ = (9\pi)^{-1} (2e/M)^{1/2} U_0^{3/2} d^{-2} \tag{1.18}$$

and

$$I_+ = \pi r_0^2 j_+ = 16^{-1} (2e/M)^{1/2} r_0^2 E_d^2 U_0^{-1/2}. \tag{1.19}$$

The danger of electrical breakdown of the accelerator gap places limits on E_d and U_0. The need to avoid having excess gas flow from the source chamber into the acceleration region limits the output radius r_0. In practice, for such systems with a small aperture the distance d cannot be less than d_0 and the calculated maximum perveance is

$$(I_+/U_0^{3/2})_{max} \approx (1/36)(2e/M)^{1/2}. \tag{1.20}$$

As an example, for hydrogen $(M = 1.6 \cdot 10^{-24}$ g) the calculated maximum perveance is roughly $5 \cdot 10^{-8}$ A/V$^{3/2}$.

These estimates do not include large distortions caused by the deviations from a Pierce system geometry related to the actual shape of the plasma boundary and to the presence of a hole in the extractor for the ions to escape. In practice so-called quasi-Pierce systems are used.

3. When the main goal is to increase the beam current, although at the cost of increasing the emittance, a third method is used for extracting a beam from the surface of a plasma emitter. Plasma passes through an aperture of diameter d_0 into the vacuum of an expansion cell, often called an expander, where a plasma emitter with an extended surface area $S = \pi D^2/4 \gg \pi d_0^2/4$ is formed. Then the width d of the layer may be considerably less than D, and the beam current is

$$I_+ = j_+ \pi D^2/4 = 36^{-1} (2e/M)^{1/2} U_0^{3/2} (D^2/d^2), \tag{1.21}$$

and the perveance

$$I_+/U_0^{3/2} = 36^{-1} (2e/M)^{1/2} (D^2/d^2) \tag{1.22}$$

is considerably greater than that given by Eq. (1.20).

This method of forming a beam, which makes it possible to increase the perveance while keeping the diameter of the aperture which controls the flow of gas into the ion acceleration region equal to d_0, was proposed in 1950 by Gabovich [1] and is used in a variety of sources. Figure 1.4 shows the location and shape of the plasma emitter as determined experimentally under different conditions for an immersion lens with a large expander. The effect on plasma focusing of a magnetic field in the expander region was first pointed out by Gabovich [1] and was also ascertained by Bacon [6] in a paper on the development of an optimum design for the expander of a duoplasmatron. Electron-optical systems of this type with an external plasma emitter have been modeled by Bortnichuk et al. [7].

4. Another route has been followed in the formation of large-perveance quasistationary beams for research on controlled thermonuclear fusion. n separate ion-optical cells (axially symmetric or slits, for which the perveance increases with the length of the cell) are used to increase the overall perveance by a factor of n. The geometry of three- or four-electrode ion-optical cells (Fig. 1.5) has been optimized [2, 8] (see Chapter 3). It is important to maintain the uniformity of the source plasma, which comes into contact with all the identical cells of the system. The perveance of hydrogen ion beams formed by hundreds of cells can be as high as 10^{-5} A/V$^{3/2}$.

Chapter 2

FOCUSING AND TRANSPORT
OF ION BEAMS

2.1 Propagation of Nonrelativistic Single-Component
Charged-Particle Beams

In a high vacuum without external fields, a dense ion beam rapidly spreads out because of the mutual repulsion of the ions. Here we note several properties of spreading single-component beams of charged particles [1]. If the length of an axially symmetric beam propagating under these conditions is considerably greater than its diameter, then the self-field of the beam is found by solving the Poisson equation

$$\frac{1}{r}\frac{\partial}{\partial r}\left(r\frac{\partial\varphi}{\partial r}\right) = -4\pi\rho. \tag{2.1}$$

When the space-charge density ρ is independent of r, the first integral of Eq. (2.1) yields the radial field

$$E(r) = 2\pi\rho r. \tag{2.2}$$

If r_0 is the beam radius, then the field strength on its surface is

$$E(r_0) = 2\pi\rho r_0 = 2I_b/(r_0 v_b), \tag{2.3}$$

and the radial potential drop is

$$\Delta\varphi = \int_0^{r_0} E(r)\,dr = I_b/v_b, \tag{2.4}$$

where v_b is the velocity of the ions in the direction of propagation of the beam and I_b is the beam current.

The radial motion of a particle on the edge obeys the equation

15

$$M\ddot{r} = F_e = 2eI_b/(r_0 v_b). \tag{2.5}$$

It is natural to assume that at the beginning of the beam path ($z = 0$) a thin lens acts on the beam so that the condition for homocentricity of the beam,

$$v_r(r) = v_b r/f, \tag{2.6}$$

is satisfied, where $v_r(r)$ is the radial velocity of a particle located a distance r from the axis and f is the focal length of the lens. For the initial conditions $t = 0$ and $r = r_0$, where r_0 is the initial beam radius, integration of Eq. (2.5) yields the following expression for $\dot{r} = -v_r(r)$:

$$(\dot{r})^2 = \frac{4eI_b}{Mv_b} \ln\frac{r}{r_0} + v_r^2(r_0). \tag{2.7}$$

This yields a minimum beam radius of

$$r_{min} = r_0 \exp[-Mv_b v_r^2(r_0)/(4eI_b)]. \tag{2.8}$$

For $r = r_{min}$ a particle stops moving radially toward the axis and begins to move away from the axis as a result of electrostatic repulsion. The second integral of Eq. (2.5) yields the shape of the converging part of the beam:

$$f\left(\frac{r_0}{r_{min}}\right) - f\left(\frac{r}{r_{min}}\right) = \left(\frac{I_b e}{M}\right)^{1/2} \frac{z}{v_b^{3/2} r_{min}} = K\frac{r_0}{r_{min}}, \tag{2.9}$$

where $f(x) = \frac{1}{2}\int_1^x \frac{dy}{\sqrt{\ln y}}$ is a standard tabulated function and K is a dimensionless parameter. For purposes of estimating the divergence of ion beams, Table 2.1 shows values of the function $f(x)$ for the most widely encountered values of x.

In the particular case where there is an initial radial velocity component, the shape of a spreading beam is given by

$$f\left(\frac{r}{r_0}\right) = \left(\frac{I_b e}{M}\right)^{1/2} \frac{z}{v_b^{3/2} r_0} = 2^{-3/4} I_b^{1/2} (M/e)^{1/4} U_0^{-3/4} r_0^{-1} z. \tag{2.10}$$

In practical units, given that $v_b = (2eU_0/M)^{1/2}$, where U_0 is the particle energy in electron volts (eV), we obtain

$$f\left(\frac{r}{r_0}\right) = 88 (M/m)^{1/4} I_b^{1/2} U_0^{-3/4} r_0^{-1} z. \tag{2.11}$$

We note that the Lorentz force determined by the magnetic field of the beam current that acts along with the force F_e and draws the beam to the axis has been neglected here. When the particle velocity is sufficiently high (comparable to the speed of light), this correction must be included.

Using Eq. (2.10), we find the shape of the beam for $z > z_{min}$ to be

$$f\left(\frac{r_0}{r_{min}}\right) = \left(\frac{I_b e}{M}\right)^{1/2} \frac{z - z_{min}}{v_b^{3/2} r_{min}}. \tag{2.12}$$

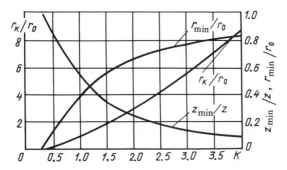

Fig. 2.1. Graphs for calculating the optimum focusing of a single-component charged particle beam.

TABLE 2.1. Values of the Function $f(x) = \dfrac{1}{2}\displaystyle\int_{1}^{x}\dfrac{dy}{\sqrt{\ln y}}$

x	$f(x)$	x	$f(x)$	x	$f(x)$
1.01	0.1	1.43	0.681	4.22	2.14
1.04	0.203	1.63	0.833	7.10	3.24
1.09	0.309	1.9	1.01	12.9	5.17
1.17	0.422	2.25	1.22	22.5	8.86
1.28	0.545	2.72	1.46	54.6	16.4

Setting $r = r_{min}$ and $f(1) = 0$ in Eq. (2.9), we obtain

$$z_{min} = f\!\left(\frac{r_0}{r_{min}}\right)\!\left(\frac{M}{I_b e}\right)^{1/2} v_b^{3/2} r_{min}\,. \tag{2.13}$$

Finally, for the diverging part of the beam we have

$$f\!\left(\frac{r_K}{r_{min}}\right) + f\!\left(\frac{r_0}{r_{min}}\right) = K\frac{r_0}{r_{min}}\,. \tag{2.14}$$

This expression makes it possible to solve an important problem: for a beam with an initial radius r_0 coming out of the plane $z = 0$ in which lies a thin lens, find the lens focal length [initial velocity $v_r(r_0)$] for which the cross section of the beam in the plane z will be minimum and determine the minimum cross section of the beam. Graphs based on Eq. (2.14) for calculating the optimum focusing of charged particle beams taking space charge into account are shown in Fig. 2.1. This graph can be used, for example, to calculate the maximum beam current passing through a cylindrical tube of length z and radius $r_K = r_0$ with optimum focusing. To do this we find from the graph that the value $r_K/r_0 = 1$ corresponds to $K_1 = 1.08$ and, taking Eq. (2.10) into account, we obtain a maximum current of

$$I_{b\,max} = 2^{3/2} (e/M)^{1/2} U_0^{3/2} (r_0/z)^2 K_1^2\,. \tag{2.15}$$

These calculations show that intense ion beams spread out rapidly in a high vacuum because of their intrinsic space charge. For example, a 1-A, 10-keV proton beam with an initial radius of 1 cm will have blown up by roughly a factor of 2 after moving a short distance just equal to the beam diameter and by a factor of 100

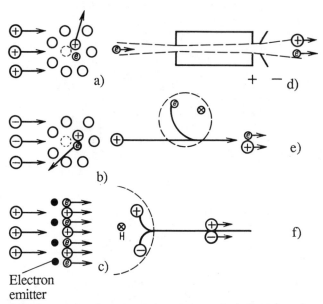

Electron
emitter

Fig. 2.2. A schematic illustration of the methods for space-charge compensation in ion beams: (a) "gas" compensation of a positive ion beam; (b) "gas" compensation of a negative ion beam; (c) formation of a synthesized ion–electron beam using a thermionic emitter; (d) formation of a synthesized beam by passing an electron beam through the ion source; (e) combining ion and electron beams with a magnetic field; and (f) formation of an ion–ion beam by combining the components in a magnetic field.

after moving 50 cm. This explains the importance of developing methods for eliminating or at least reducing the effect of space charge. The most natural way of solving this problem is to introduce charges of the opposite sign into an ion beam (Fig. 2.2).

2.2 Gas Compensation of Stable Positive or Negative Ion Beams

We now consider in more detail the self-compensation or gas compensation of ion beams (Fig. 2.2a,b) passing through the residual gas or through a gas specially introduced into the ion beam duct [9, 10]. Under these conditions $v_e^+ = n_b v_b n_a \sigma_e$ electrons are formed per cm^3 per unit time in the medium traversed by a positive ion beam, where n_a is the neutral atom density and σ_e is the cross section for electron production. For a negative ion beam the corresponding quantity is $v_e^- = n_b v_b n_a (\sigma_e + \sigma_{\bar{1},0})$, where $\sigma_{\bar{1},0}$ is the cross section for electron detachment. Simultaneously with the electrons, $v_{pi} = n_b v_b n_a \sigma_{pi}$ slow positive ions are produced by these beams, where σ_{pi} is the cross section for production of these ions through ionization and charge exchange.

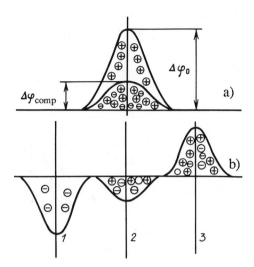

Fig. 2.3. The radial distributions of the potential in a positive ion beam before and after compensation (a) and in a negative ion beam for various gas pressures in the ion beam duct (b). The potential distributions (1) in high vacuum, (2) at high pressure, and (3) when the gas pressure is raised further and the sign of the potential reverses.

Equation (2.4) shows that in an uncompensated positive ion beam an electron potential "well" is formed with a depth of

$$\Delta\varphi = I_b/v_b = en_b\pi r_0^2. \tag{2.16}$$

The slow ions which are produced are expelled from the beam while the electrons accumulate in a potential "gutter" whose initial depth is given by Eq. (2.16). A reduction in $\Delta\varphi$ then takes place; i.e., the beam is compensated. Because the average velocities of the slow ions and electrons are related by the inequality $\bar{v}_{pi} \ll \bar{v}_e$, while $v_{pi} \geq v_e$, gas compensation of a freely propagating beam of positive ions can only reduce the electric field responsible for expansion of the beam. The field cannot change sign in this case (Fig. 2.3a).

Coulomb collisions of beam ions with electrons. In early work [11] the equilibrium value of $\Delta\varphi_{comp}$ in a stable compensated ion beam was evaluated using only the balance equation for the electrons

$$v_e\beta\pi r_0^2 = 2\pi r_0 (n_e v_e/4) \exp[-e\Delta\varphi_{comp}/(\chi T_e)], \tag{2.17}$$

where β is the fraction of the electrons which are captured by the equilibrium residual potential well of the beam since their energy is not high enough for them to escape. Here T_e is the temperature of the compensating electron gas. The expression $\Delta\varphi_{comp} = (\chi T_e/e)\ln\varphi$ derived from Eq. (2.17), however, only gives the potential drop in terms of another unknown quantity, T_e. The basis of this mechanism for maintaining a stationary state is energy exchange between electrons captured by the well and the loss of electrons from the well through the "tail" of a Maxwellian velocity distribution. This is inconceivable without an additional source for the energy transferred to the electron gas. It would seem that the space charge of the beam must be fully compensated by the charge of the accumulated slow electrons. In any case, this problem can be solved correctly only when the energy balance equation is used along with the particle balance equation. It has been shown that the ion beam itself serves as a supplementary energy source [9, 10]. A steady state with a finite value of $\Delta\varphi_{comp}$ is maintained because, during Coulomb collisions of beam ions with electrons captured in the potential well, the latter acquire additional energy and leave the beam. Under ordinary conditions these Coulomb collisions are extremely rare, but the power that must be supplied to a compensated beam in order to drive all the captured electrons out of the well is also very small and is given by

$$E = L \int_0^r 2\pi\xi d\xi \int_0^{e\varphi(\xi)} \psi(\epsilon)(e\varphi - \epsilon)d\epsilon, \tag{2.18}$$

where L is the length of the beam. The function $\psi(\xi)$ characterizes the energy distribution of the electrons that are produced $[\int_0^\infty \psi(\epsilon)d\epsilon = 1]$ and is approximated by the formula $\psi(\epsilon) \propto 1/(\epsilon + \epsilon_i)^2$ derived from the classical theory of Thomson ionization ($\epsilon_i = e\varphi_i$ is the ionization energy of an atom) and the shape of the potential well is given by the function

$$\varphi(\xi) = \Delta\varphi_{comp}(1 - \xi^2/r_0^2). \tag{2.19}$$

The power transferred by the ion beam to the electron gas is

$$Q_{Coul} = \frac{an_b n_e e^4 \pi r_0^2 L}{mv_b}, \tag{2.20}$$

where $\alpha = 4\pi \ln [m^{3/2}v_b/(1.78\pi^{1/2}n_e^{1/2}e^3)]$.

Before obtaining an expression for the equilibrium value of $\Delta\varphi_{comp}$, we should note that for relatively low gas pressures, when the density of atoms is considerably lower than

$$n_{ao} = 2\bar{v}_{pi}/(r_0 v_b \sigma_{pi}), \tag{2.21}$$

the density n_{pi} of slow ions is much less than n_b and the actual ion beam plasma can be modeled by an ion beam which is compensated by the electrons so that the quasineutrality condition

$$n_e \lesssim n_b \tag{2.22}$$

is satisfied. As the gas pressure is raised, condition (2.22) is replaced by

$$n_e \lesssim n_b + n_{pi}, \tag{2.23}$$

in which the density of slow ions is approximated by

$$n_{pi} = n_b v_b n_a \, q_i r_0 / (2 \bar{v}_{pi}),$$ (2.24)

obtained, as was Eq. (2.21), from the appropriate particle balance. In the second case the system is an ion beam propagating through a two-component plasma produced by the beam itself.

Residual potential of a stable compensated positive ion beam. $\Delta\varphi_{comp}$ is obtained from the equality $E = Q_{Coul}$ and Eqs. (2.23) and (2.24). $\bar{v}_{pi}$ depends both on the initial energy of the ions and on the field created by the beam. With good compensation, when $\Delta\varphi_{comp} < \varphi_i$, the depth of the potential well of a stable ion beam can be obtained using the relatively simple expression

$$\Delta\varphi_{comp} = (3a)^{1/2} e (M_b/m)^{1/2} (\varphi_1/\varphi_b)^{1/2} \left(\frac{1}{n_a \sigma_e} + \frac{v_b \sigma_{pi} r_0}{2 \bar{v}_{pi} \sigma_e} \right)^{1/2} n_b^{1/2}.$$ (2.25)

Measurements of the radial potential drop in a long beam [9,10] have made it possible to compare the theoretical dependences $\Delta\varphi_{comp} = \Phi(r_0, n_a, n_b)$ with experimental data and to ascertain that they agree. The predicted self-decompensation of a beam by Coulomb collisions between the ions and compensating electrons has, therefore, been confirmed experimentally.

Self-focusing of negative ion beams. Experimental studies of the dependence of the potential in a negative ion beam on the gas pressure have shown that at low pressures the potential remains negative, despite the possibility in principle that slow positive ions produced by the beam may accumulate there. At some pressure p_0 the potential is equal to zero and for $p > p_0$ it changes sign (Fig. 2.3b). This circumstance is naturally of interest because it indicates that self-focusing or gas focusing of negative ion beams that have been overcompensated by positive ions may be possible. The density of gas atoms corresponding to p_0 is determined from the balance condition for formation and loss of positive ions when $n_b \approx n_{pi}$ and is equal to n_{a0} given by Eq. (2.21). Proceeding from the idea that the electrons leave an overcompensated negative ion beam because they undergo Coulomb collisions with the negative ions, it is possible to estimate the order of the potential drop in such a beam when $p > p_0$:

$$\Delta\varphi^-_{comp} = a^{1/2} e (M/m)^{1/2} (\varphi_i/\varphi_b)^{1/2} \frac{1}{n_{a0}^{1/2} (\sigma_{pi} + \sigma_{\bar{1},0})^{1/2}} \left(1 - \frac{n_{a0}}{n_a} \right)^{1/2} n_b^{1/2}.$$ (2.26)

Equation (2.26) is qualitatively confirmed by experiment [9, 10].

One reason for the limited use of self-focusing in negative ion beams is that even for $p = p_0$ the mean free path for detachment

$$\lambda_{\bar{1},0} = 1/(n_{a0} \sigma_{\bar{1},0}) = r_0 v_b \sigma_{pi} / (2 \bar{v}_{pi} \sigma_{\bar{1},0})$$ (2.27)

is small; i.e., the negative ions are converted into fast neutral atoms after moving a relatively short distance.

2.3. Spreading of Compensated Positive Ion Beams and Its Effect on Focusing

For examining the spreading of compensated positive ion beams by residual space charge Gabovich [12] has proposed using the Kapchinskii equation and representing the beam envelope by the equation

$$d^2r/dz^2 - \mathscr{E}^2/r^3 - \Delta\varphi_{comp}/\varphi_b r = 0. \tag{2.28}$$

Here $\mathscr{E} = (2^{3/2}R_{00}/v_b)(\chi T_b/M_b)^{1/2}$ is the emittance, R_{00} is the beam radius in the region where ions are collected from the emitter surface, and $\Delta\varphi_{comp}$ is given by Eq. (2.25). At relatively low gas pressures, when $(n_a\sigma_e)^{-1} \gg -v_b\sigma_{pi}r_0/(2\bar{v}_{pi})$, Eq. (2.25) is simplified and integrating Eq. (2.28) yields the desired relationship between the reduced beam radius $\rho = r/r_0$ and length $z_{comp}' = (z_{comp}/r_0)(2\Delta\varphi_{comp0}/\varphi_b)^{1/2}$, where z_{comp} is the actual length of the compensated beam, r_0 is its initial radius, and $\Delta\varphi_{comp0} = \Delta\varphi_{comp}r/r_0 = (3\alpha)^{1/2}e(M_b/m)^{1/2}(\varphi_i/\varphi_b)^{1/2}(n_a\sigma_e)^{-1}[I_b/(\pi r_0^2 e v_b)]^{1/2}$. The function $z_{comp}' = f(\rho)$ has the form

$$z'_{comp} = \frac{[(1+a)\rho^2 - \rho - a]^{1/2}}{1+a}$$
$$+ \frac{1}{2(1+a)^{3/2}} \ln \frac{2(1+a)^{1/2}[(1+a)\rho^2 - \rho - a] + 2(1+a)\rho - 1}{1+2a}, \tag{2.29}$$

where the parameter which characterizes the thermal speeds is

$$a = \frac{\mathscr{E}^2\varphi_b}{2r_0^2\Delta\varphi_{comp\,0}} = \frac{2R_{00}^2\chi T_b}{r_0^2 e\Delta\varphi_{comp\,0}}.$$

Figure 2.4 shows profiles of a spread-out compensated beam obtained from Eq. (2.29) for two values of a.

By comparing Eq. (2.10) for a single-component beam with a particle temperature of zero and Eq. (2.29) for $a = 0$, it is possible to compare the spreading of uncompensated and compensated beams. This yields the ratio

$$\frac{z_{comp\,0}}{z} = F(\rho)\left(\frac{\Delta\varphi}{\Delta\varphi_{comp\,0}}\right)^{1/2}, \tag{2.30}$$

where $F(\rho)$ is a function which is close to unity when ρ is not very large and $\Delta\varphi$ is the radial potential drop in a single-component beam given by Eq. (2.4). The ratio $\Delta\varphi/\Delta\varphi_{comp0}$ characterizes the compensation effect. It follows, for example, from

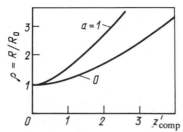

Fig. 2.4. Profiles of spread-out compensated ion beams for two values of the parameter a.

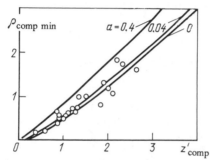

Fig. 2.5. The dependence of the minimum radius of a compensated beam on its reduced length. (The points are experimental data.)

Eq. (2.30) that a 100-fold reduction in the radial potential drop during compensation of an ion beam leads to a 10-fold increase in the length of the beam over which the given spreading (given ρ) occurs.

A comparison of Eqs. (2.10) and (2.29) leads to yet another important conclusion. It follows from Eq. (2.10) that for a given energy and length of a single-component beam its spread is uniquely determined by the initial current density. As opposed to this, the spreading of a stable compensated beam is determined by the initial beam radius as well as by the initial current density. For a given current density the spread decreases as the initial radius or overall beam current increases. This can be seen in Fig. 2.4, where the growth in ρ with increasing z_{comp}' is apparent.

It was shown above that, using the curves of Fig. 2.1 obtained from Eq. (2.14), it is possible to find the minimum cross section of a single-component beam in a given plane $z = 0$. The analogous problem has also been solved for compensated beams [13, 14]. For beam propagation through a relatively low pressure gaseous atmosphere, the problem reduces to the joint solution of the equations

$$\int_{\rho_m}^{1} \frac{\rho d\rho}{\sqrt{c\rho^2 - \rho - a}} + \int_{\rho_m}^{\rho_{comp}} \frac{\rho d\rho}{\sqrt{c\rho^2 - \rho - a}} = \int_{0}^{z_{comp}'} dz \qquad (2.31)$$

and

$$\partial\rho_{comp} \, |\partial f|_{z\,comp} = 0. \tag{2.32}$$

Here f is the focal length of the lens; $\rho_m = [1 + (1 + 4ac)^{1/2}]/2c$ is the beam radius at the crossover (where $\partial\rho/\partial z = 0$) obtained by integrating Eq. (2.28) for a given f; $c = 1 + a$; and ρ_{comp} is the beam radius in the plane z_{comp}, whose minimum value $\rho_{comp\,min}$ is to be determined (it corresponds to some optimum value of f).

Solving Eqs. (2.31) and (2.32) jointly on a computer yields the desired dependence of the minimum radius of the compensated beam, $\rho_{comp\,min}$, on the parameter z_{comp}' (Fig. 2.5). This figure can be used to evaluate the agreement between the computational results and the corresponding experimental data.

2.4. The Effect of the Self-Magnetic Field of a Compensated Ion Beam on Beam Transport

It was shown above that an increase in the radius or overall current of a compensated ion beam with a given density leads to a reduction in the radial electric field and, therefore, to reduced spreading. Under these conditions it may seem unnecessary to take the magnetic field of the beam current [9], which causes pinching of the beam, into account. Including the Lorentz force acting on particles at the periphery of the beam, $F_H = ev_bH/c = 2ev_bI_b/c^2r$, the Kapchinskii equation takes the form

$$d^2r/dz^2 - \varepsilon^2/r^3 - \Delta\varphi_{comp} \, /(\varphi_b r) + v_b I_b/(c^2 r \varphi_b) = 0. \tag{2.33}$$

Comparing the second and last terms in this equation gives an estimate of the beam current at which the influence of the beam's self-magnetic field on spreading exceeds that of the emittance. For protons with $R_{00} = r$ this critical current satisfies the inequality

$$I_{b\,cr} > 100 T_b/\varphi_b^{1/2}, \tag{2.34}$$

where T_b is in eV, φ_b is in kV, and $I_{b\,cr}$ is in A. If, on the other hand, we neglect the emittance, then the beam current at which the effect of its magnetic field predominates over the effect of the residual electric field obeys the inequality

$$I_{b\,cr} > 20\Delta\varphi_{comp} \, /\varphi_b^{1/2}, \tag{2.35}$$

where $\Delta\varphi_{comp}$ is the experimentally determined residual radial potential drop (V), φ_b is in kV, and $I_{b\,cr}$ is in A.

The influence of the self-magnetic field of a beam on beam transport has been specially examined by Cottrell [15], who obtained a formula for $I_{b\,cr}$ that is the same, except for the coefficient, as Eq. (2.34). He neglected the residual electric field. On the other hand, a comparison of Eqs. (2.34) and (2.35), as well as the known orders of magnitude of $\Delta \varphi_{comp}$ and T_b, shows that in a number of cases it is not correct to neglect the electric field.

2.5 Synthesized Beams

Under high-vacuum conditions gas compensation is clearly ineffective and, without resorting to methods that involve a local enhancement in the pressure [16], a beam can be compensated by other means, specifically, by introducing particles of the opposite charge from outside. Synthesized beams, consisting of interpenetrating ion and electron beams moving in the same direction, are extensively used. When the densities and velocities of the components are equal, this method compensates the space charge and ensures that the total current is zero. This latter condition is necessary, for example, when an ion beam is used for rocket propulsion. In order to keep a propelled isolated body from accumulating charge, it must emit the same amount of electrons as ions. Otherwise the body would acquire a negative potential and cease to emit ions.

Ion–electron synthesized beams. A compensating electron beam can be created with the aid of thermal electron emitters placed in the path of the ion beam (Fig. 2.2c). It can also be produced by passing a shaped, accelerated electron beam through the ion source (Fig. 2.2d) or by using a magnetic field to combine an electron beam with an ion beam (Fig. 2.2e, etc.). In order that the velocities of the ions and electrons in a synthesized beam be equal regardless of how the beam is produced, the potential differences that determine the energies of these particles must be proportional to the ratio of the corresponding masses, M_b/m. An ideal system consisting of cold electron and ion beams with equal charge densities and velocities (parallel to the propagation direction) at any point should not spread since there are no electric fields or radial velocities in it. A real synthesized beam does not behave this way since at its initial cross section the charged particles of different signs have thermal velocities in addition to directed velocities, and their densities, although the same, are not constant over the beam cross section. Numerical solutions of the equations that describe the propagation of synthesized beams have yielded some interesting characteristics of the propagation process [9, 10]. When the numbers of positive and negative charges per unit length of the beam are kept equal, part of the electrons leave the ion core of the beam, penetrate into the space outside the beam, and form an electron "shell." Because of charge separation, an electric field develops which confines the shell but simultaneously makes the ion component spread. A longitudinal potential distribution also develops, and under these conditions radial oscillations of the electron component lead to a growth in the transverse components of the phase volume of the beam, which is equivalent to an increase in the electron temperature. Therefore, a real synthesized ion–electron beam has a radial electric field which causes spreading of the beam. The radial potential drop in such a steady-state synthesized beam is proportional to $T_e^{1/2}$, where T_e is the temperature of the electron component, which is considerably greater than that of the electron source or neutralizer.

An ion beam can also be compensated by a dense plasma source such as a hot tungsten surface in cesium vapor.

Ion bombardment and the potential of dielectric surfaces. A poorly conducting or conducting, but insulated, body develops a stationary potential when the total current of particles incident on and leaving the body is equal to zero. Therefore, the commercial use of positive ion beams for materials processing in high vacuum is often difficult because of charging of the surface of the material and the appearance on it of a positive potential as high as the ion source potential, so that ions no longer reach the surface or arrive with reduced energy. Secondary electron emission from the target merely facilitates the establishment of a high positive potential. Only when the pressure of the residual gas is fairly high can the source potential be significantly reduced by the arrival on the surface of electrons from the ion-beam plasma.

A beam–dielectric system behaves quite differently during bombardment of a target by negative ions [17] with energies such that the ion–electron emission coefficient $\gamma_- > 1$. In this case, the influx of ions is compensated by loss of part of the electrons knocked out of the surface, and even in a high vacuum a small positive potential for which the total particle current will be zero can be established and ions will fall on the target without significant changes in their initial energy. Thus, negative particle beams are sometimes attractive for materials processing. We also note the good focusability of these beams owing to the fact that a small positive potential on an insulated body promotes the accumulation of slow positive ions in the beam which effectively compensate the negative space charge.

Ion–ion synthesized beams. Using negative ions is not the only method of removing charge under high vacuum conditions. Another method is to use a synthesized ion–electron [18] or ion–ion beam with a zero total particle current (Fig. 2.2f). A synthesized ion–ion plasma was first created at the Institute of Physics of the Ukrainian Academy of Sciences in 1970 [19, 20].

Influence of transverse magnetic fields on synthesized beams. In the path of propagating beams there is a transverse (relative to the propagation direction) magnetic field which affects the transport of synthesized beams [21]. The currents of both components of a synthesized beam with equal densities $n_\pm$ and velocities $v_\pm$ are given by

$$I_\pm = en_\pm v_\pm l\delta, \tag{2.36}$$

where l is the height of the beam and δ is its width. (For simplicity we consider a ribbon beam.) During propagation in a region with a transverse magnetic field the Lorentz force

$$F_\pm = (e/c) [v_\pm H] \tag{2.37}$$

causes polarization of the charges. In the case of complete separation of the latter the resulting electric field acts on unit charge with a force

$$F_e = 2\pi e^2 n_\pm \delta. \tag{2.38}$$

The condition for rectilinear propagation of a two-component beam perpendicular to a magnetic field has the form

$$F_e > ev_\pm H/c \quad \text{or} \quad I_\pm > v_\pm^2 Hl/(2\pi c). \tag{2.39}$$

We note that polarization of ion–electron beams in a magnetic field has been studied experimentally [19]. It was found that for a sufficiently high field H a synthesized beam is completely separated into its components and an insulated target, on which only positive ions fall under these conditions, acquires a high positive potential that "blocks" the beam. The critical magnetic field is in agreement with Eq. (2.39).

Transport of an ion beam can be facilitated by combining it with an electron beam of equal space-charge density but with a large current. The intrinsic magnetic field of the compensated electron beam causes self-compression, and the interrelation of the positive and negative space charges causes contraction of the positive ion beam as well [22].

2.6 Collective Processes in Ion-Beam Plasmas and Their Effect on Transport of Compensated Beams

Ion-beam plasmas. A compensated ion beam is affected both by ion–electron Coulomb collisions and by the collective processes that develop in it. A positive ion beam that is compensated by electrons can be viewed as an ion-beam plasma when the quasineutrality condition (2.22) is supplemented by the inequality

$$d_e = \left(\frac{\chi T_e}{4\pi n_e e^2}\right)^{1/2} \ll r, \tag{2.40}$$

which means that the Debye screening length is considerably less than the beam radius. The inequality (2.40) can be written in the form $I_b/v_b \gg \chi T_e/e$, i.e., $\Delta\varphi_b = I_b/v_b \gg \Delta\varphi_{comp}$. The latter form means that a beam with a substantial compensation effect $\Delta\varphi_0/\Delta\varphi_{comp}$ is an ion-beam plasma. As the gas pressure is raised, a two-component ion-beam plasma becomes a three-component plasma that obeys the quasineutrality condition (2.23). For example, when beams with a current $I = 10$ A pass through a charge exchange chamber, a plasma is formed at the entrance of the latter with an ion density $n_{pi} = 5\cdot10^{10}$-$5\cdot10^{11}$ cm^{-3} and $T_e \approx 1$ eV. The density falls with increasing distance from the ion source, while the electron temperature increases [8]. A temperature T_{pi} can be assigned to the slow ions only arbitrarily.

Electron plasma (Langmuir) oscillations. When the electrons in a plasma are displaced, the quasineutrality of the plasma is disrupted and an electric field develops. A restoring force $F = -eE = -4\pi e^2 n_e x$ acts on each of the displaced electrons and the density difference $n_+ - n_e$ gradually decreases to zero. The electrons, however, continue to move because of inertia, so that the quasineutrality is again disrupted and this process repeats itself periodically. Equating the force F to the force of inertia yields the equation $m\ddot{x} + 4\pi e^2 n_e x = 0$, which gives the frequency of the electron plasma (Langmuir) oscillations,

$$\omega_e = (4\pi n_e e^2/m)^{1/2}. \tag{2.41}$$

The high-frequency electron oscillations take place against the uniform background of the relatively motionless (because of their large mass) positive ions.

When the electrons have thermal motion, the frequency of their oscillations depends on the wavelength or wave number $k = 2\pi/\lambda$ as

$$\omega^2 = \omega_e^2 + k^2\, 3\chi\, T_e/m. \tag{2.42}$$

Similar electron plasma oscillations occur in ion-beam plasmas. Electron oscillations are excited because in the presence of a beam the number of particles that are ahead of a wave with a given phase velocity $v_{ph} = \omega/k < v_b$ is greater than the number that are behind and the particles ahead of the wave give energy to the wave. The pumping length for the oscillations is $L \sim v_b/\gamma$, where γ is the temporal growth rate, which is given by

$$\gamma = 3^{1/2}\, 2^{-4/3}\, \omega_e\, (m/M_b)^{2/3} \tag{2.43}$$

for a low-pressure gas and is obtained from the dispersion relation for a plasma with a beam propagating in it:

$$1 - \omega_e^2/\omega^2 - \alpha\omega_e^2/(\omega - k_z\, v_b)^2 = 0. \tag{2.44}$$

Here $\alpha = n_b m/(n_e M_b) \ll 1$.

The principal nonlinear effect that limits the amplitude of the plasma oscillations in this plasma is trapping of the compensating electrons by the wave field [9, 10]. Moving at a speed close to the ion beam velocity v_b, the wave "rakes up" and entrains the cold compensating electrons when its amplitude reaches

$$e\,\tilde{\varphi}_{max} = mv_b^2/4e. \tag{2.45}$$

This process involves a loss of energy by the wave and limits its amplitude. Electron plasma oscillations are also excited in analogous beams of negative ions. Of course, this occurs only for $p > p_0$, when the change in sign of the potential from negative to positive is accompanied by an accumulation of electrons in the beam [23].

While Coulomb collisions of an ion beam with electrons cause ejection of the latter predominantly in the radial direction, when collective oscillations are excited the electrons are predominantly accelerated in the direction of propagation of the beam. Direct experimental proof of decompensation owing to excitation of electron oscillations has been obtained in a study of helium ion beams with a current of 12 mA, an energy of 120 keV, and an ion density $n_b \approx 10^8$ cm^{-3} moving inside a 200-

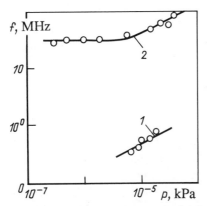

Fig. 2.6. The dependence of the frequencies of ion (1) and electron (2) oscillations observed in a compensated ion beam on the gas pressure.

cm-long vessel with a residual gas pressure of 10^{-3}-10^{-1} Pa [24]. Waves of different amplitudes were excited with a three-grid beam modulator. In the absence of external modulation, electron plasma oscillations were spontaneously excited in the system and grew exponentially along the beam. The amplitude of these spontaneous oscillations, however, was less than 1 V even at the end of the system, so they had no noticeable effect on the beam potential resulting from Coulomb collisions (up to 13 V). As the amplitude of the modulation in the ion velocity was raised, the variable $\tilde{\varphi}$ and constant φ potentials in the beam were observed to increase. For a certain modulation amplitude $\tilde{\varphi}$ and φ stopped increasing. All of this is explained by the above-mentioned capture of previously "motionless" compensating electrons by the wave. It may be concluded that for sufficiently large amplitude oscillations the increase in $\tilde{\varphi}$ and φ (decompensation by electron oscillations) is mainly determined by the velocity of the ion beam.

Ion oscillations. Besides electron oscillations against the uniform ion background, ion plasma (Langmuir) waves with a frequency of $\omega_{pi} = (4\pi e^2 n_+/M_{pi})^{1/2}$ are observed against the uniform electron background. The conditions for this exist in a nonisothermal plasma with cold ion and hot electron components ($T_e/T_{pi} \gg 1$), where because of their high temperature the electrons are "smeared out" in space regardless of the potential profile of the ion oscillations. Ion-acoustic oscillations, whose frequency dependence on the wave number $\omega = k(\chi T_e/M_{pi})^{1/2}$ is analogous to that for ordinary sound waves, belong to the same low-frequency branch as ion plasma waves. Unlike sound waves, which propagate in an atmosphere of neutral particles, ion-acoustic waves propagate in a plasma where the ion and electron motions are coupled. In this case the electrons are no longer an independent background; the restoring force is determined by the electron pressure $n_\chi T_e$ and the inertia is from the ions. Thus, the ion sound speed $c_s = \omega/k$ is determined by the electron temperature and the mass of the heavier particles, the ions. The pumping of ion-acoustic waves requires a resonance beam whose velocity equals the phase velocity of these waves. Therefore, a slow ion beam with a velocity on the order of c_s excites low-frequency longitudinal oscillations in a plasma [25]. On the other hand, in an ion-beam plasma a fast ion beam excites oscillations which do not propagate along the beam, but at an angle θ to the beam such that $v_b \cos \theta \approx c_s$. Since usually $c_s/v_b \ll 1$, the angle θ is close to $\pi/2$. A detailed theoretical analysis [25] shows that an ordinary ion beam excites "oblique" waves which propagate almost perpendicular to the ion beam with a frequency

$$\omega_k = \omega_{pi} / [1 + 1/(k^2 \, d_e^2)]^{1/2} \qquad\qquad (2.46)$$

and a maximum growth rate of $\gamma = 3^{1/2} \cdot 2^{-4/3} \omega_k (\omega_b/\omega_{pi})^{2/3}$. When $k^2 d_e^2 \ll 1$, ion-acoustic waves are excited, and when $k^2 d_e^2 \gg 1$, ion plasma waves are excited. Oblique waves of this type were observed in 1973-1975 at the Institute of Physics of the Ukrainian Academy of Sciences [26, 27]. An experimental dependence $\omega = f(k)$ that agreed well with theory [28] was obtained by modulating the ion beam at different frequencies and observing the corresponding phases at different distances from the beam axis. It also seems that measuring the limiting frequency of the oscillations (that is, the ion plasma frequency) at large k is a method for determining the slow ion density.

The regions for exciting electron and ion oscillations are shown in Fig. 2.6 [26], which shows the frequencies of the high- and low-frequency oscillations excited in an ion-beam plasma as functions of the gas pressure. By lowering or raising the gas pressure it is possible to reach conditions under which only electron oscillations are excited. This is natural, since in the first case electrons are still accumulated from the gas and their density is negligible. At relatively high pressures, when damping of the oscillations becomes significant because of collisions of charged particles with atoms, only electron oscillations with a large growth rate are observed.

Ion oscillations in negative ion beams. Because good conditions exist in negative ion beams for the accumulation of positive ions which also compensate the space charge, the excitation of ion oscillations is even more marked than the analogous process in positive ion beams. The range of pressures at which ion oscillations are observed in negative ion beams is considerably wider than for the positive ion beams illustrated in Fig. 2.6. The most interesting feature of ion oscillations in compensated negative ion beams, however, is that at low gas pressures, when the potential in the beam is negative, electrons are expelled from it, their density is negligible, and they cannot significantly affect the ion oscillations, in particular, the amplitude of the resulting fields. Fluctuating fields with a characteristic frequency of ω_{pi} and an amplitude of tens of percent of the field in an uncompensated beam have been observed in dense negative ion beams compensated by positive ions. These fields arise because the beam excites transverse ion oscillations and may be a serious obstacle to the propagation of dense negative ion beams [29], whose transport is presently of interest in connection with the development of fast neutral particle injectors based on negative ion beams.

Transverse ion bunching. We now consider effects that limit the amplitude of ion oscillations in ion-beam plasmas [9, 10]. If we specify a periodic perturbation in the transverse ion velocity $v_x = v_0 \sin (kx) \sin (\omega t)$, then for a low

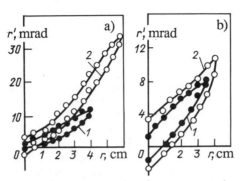

Fig. 2.7. Phase portraits of a compensated ion beam before (1) and during (2) modulation at a frequency of 20 MHz, the frequency of electron (a) and ion (b) oscillations.

gas pressure, when the plasma consists only of beam ions and electrons, the trajectories of the ions collected in the separate beams intersect in some plane $z = z_1$. This is caused by a transverse bunching of the ions; i.e., ions with large initial velocities catch up with ions that previously passed through the same zone of the modulator but acquired a lower transverse velocity. When the plasma pressure is relatively high, there are more slow ions and transverse ion oscillations are excited. Under these conditions the beam interacts with the ion oscillations and the transverse velocities of the beam particles change over the entire propagation path, and not just in the modulator. Some particles are continuously accelerated and others are slowed down. As a result, the trajectories of the beam ions intersect earlier, in a plane $z_2 < z_1$. This spatial–temporal focusing of an ion beam, i.e., bunching and the intersection of the ion trajectories, is a mechanism for limiting the amplitude of the ion oscillations. This effect is analogous to the shortening of the distance to a phase focus when an electron beam interacts with longitudinal electron plasma oscillations.

Effect of collective oscillations on the phase volume of beams. We now examine the effect of collective oscillations on the propagation of compensated ion beams. It has already been pointed out that one important parameter of an ion beam is its normalized emittance (transverse phase volume), which under ideal conditions should not change along the entire beam length. For a number of reasons, however, the actual phase volume does not remain constant. One such reason is the excitation of electron and ion oscillations in compensated ion beams. The variable electric fields in these oscillations cause spreading in the transverse ion velocities and, thus, lead to an increase in the normalized emittance that is actually measured (averaged over the period of the oscillations). The fact that transverse ion oscillations are excited in compensated positive ion beams means that variable electric fields exist which increase the spread in the transverse velocities of the ions. Electron oscillations, on the other hand, are longitudinal, and in a radially unbounded system they cannot cause any change in the radial velocities. In real, radially bounded beams, however, these fields also have transverse components. Knowing that the amplitude of the wave is limited owing to capture of electrons by the wave, it is possible in this case to evaluate the maximum scatter in the ion velocities.

Without dwelling on specific calculations [30], we shall simply consider some so-called phase portraits of a compensated ion beam (Fig. 2.7) that have been obtained experimentally by modulating the longitudinal velocity of the ions. In the case of an unmodulated beam the spontaneously excited oscillations could be neglected. It follows from this figure that excitation of electron oscillations causes an increase in the phase volume and, simultaneously, a noticeable spreading of the beam. Calculations show that the observed spreading cannot be explained solely by an increase in the transverse ion velocities (i.e., be attributed to an increase in the phase volume). The additional spreading is related to the fact that excitation of electron oscillations also leads to heating of the electron gas that facilitates decompensation of the ion beam, that is, to an increase in the electrostatic field strength. The heating of the electron gas and the decompensation of the beam during excitation of electron oscillations were confirmed directly in independent experiments. Figure 2.7b shows the phase portraits of a compensated ion beam when it is not modulated and when it is modulated at a frequency of 20 MHz, close to the ion plasma (Langmuir) frequency. Here the modulator provided a periodic (in time and in the transverse direction) perturbation of the velocity. The increased phase volume and the small observed spreading of the beam are explained by excitation of ion plasma oscillations that lead to a greater spread in the transverse velocities of the ions. Calculations show that as the energy of an ion beam is raised, the first-order increase in the phase volume is caused by an increased contribution from electron plasma oscillations.

The effect of collective oscillations excited in a plasma by an ion beam on the beam propagation may be illustrated by the following example. As it passes through a gas target a beam forms a nonisothermal plasma and the ion oscillations discussed above are excited and propagate almost normal to the beam. Pistunovich et al. [31] have observed anomalous scattering of a dense ion current (above 10 A/cm^2) directly associated with the excitation of ion oscillations. It is also important, however, that the ion beam simultaneously excited electron oscillations, heated the plasma electrons, and, since it thereby kept the plasma nonisothermal, promoted the excitation of ion oscillations and anomalous scattering. An understanding of the nature of anomalous scattering also made it possible to find means for fighting it. As might be expected, introducing hot ions to the plasma from another source, in order to equilibrate the ion and electron temperatures, stabilized the ion instability and reduced the scattering.

2.7. Compensation of Fluctuating Ion Beams and Ion Clusters (Dynamic Decompensation of Ion Beams)

The two dynamic decompensation regimes. The parameters of a real ion-source plasma and, therefore, the beam current oscillate with an amplitude and frequency that are related to plasma instabilities, even when all the power supplies are stabilized. Let the beam current have the simple time dependence $I_b = I_{b0}[1 + \alpha \cos (\omega t)]$. Dynamic decompensation occurs when the number of electrons formed per cm^3 during the half period of the oscillations is considerably smaller than the amplitude of the ion density oscillations in the beam, i.e., when

$$n_{b0}\, v_b\, n_a\, \sigma_e / (2f) \ll a\, n_{b0}. \qquad (2.47)$$

Significant fluctuating and dc potentials may then develop in the ion beam region relative to the vessel walls. Two dynamic decompensation regimes have been observed [32]. One of them occurs when the frequency f is not very high, the wavelength of the space-charge wave is greater than the length L of the device (i.e., the beam length), electron capture by the fields of the space-charge waves can be neglected, and the excess electrons are lost during the half period with the minimum potential. Then the variable potential in the beam is

$$\tilde{\varphi} \approx a\, I_b / v_b, \qquad (2.48)$$

and the minimum constant potential differs from that determined by Coulomb collisions in a stable beam in that the term $(n_a \sigma_e)^{-1/2}$ of Eq. (2.25) acquires a factor $(1 - \alpha)$. Equation (2.48) is also applicable to negative ion beams.

Beams behave differently at rather higher frequencies $f > v_b/L$. In this case electrons can be captured by the field of the space-charge wave and be ejected from the beam both by Coulomb collisions and by entrainment of captured electrons. Because of this the value of φ_{min} must be greater than at low frequencies. In addition, the variable space charge of the beam must be partially compensated by the plasma electrons, so that $\tilde{\varphi}$ should be lower than in the previous case. This has all been confirmed in experiments with a helium ion beam (current $I_b = 20$ mA, energy 25 keV, length $L = 200$ cm) that was artificially modulated over a frequency range from hundreds of kHz to 20 MHz. At low gas pressures (10^{-2} Pa) $\tilde{\varphi}$ and φ_{min} were tens of volts. As the pressure was brought up to 10^{-2} Pa the effect of dynamic decompensation was significantly reduced.

Nezlin and Zharinov [33, 34] have examined beams in transverse magnetic fields and shown how fluctuations in the plasma density cause periodic changes in the shape and location of the plasma emitter boundary and lead to enhanced dynamic decompensation. By reducing the amplitude of the oscillations in ion sources and choosing optimum operating conditions for the ion-optical system (including the plasma boundary), it was possible to improve the beam transport, in particular, to increase the productivity of electromagnetic isotope separators.

Compensation of ion clusters. The compensation of ion clusters is of independent interest [35]. When an isolated short cluster of positive ions moves inside a metal ion duct, it produces electrons and slow positive ions along its path. The moving potential well formed by the cluster may capture the electrons that compensate the space charge of the cluster. The space-charge potential may, in principle, fall to a value such that electrons are still captured by the cluster: $e\varphi_{min} \geq m v_b^2/2$. This is often several volts. However, in order for compensation to occur down to this minimum potential yet another condition must be met. A sufficient number of electrons can be generated if the beam has moved a distance $L \geq 1/(n_a \sigma_e)$.

Under ordinary experimental conditions the density of atoms is low and the path L is so large that the cluster spreads out before it is compensated. Gas compensation of positive ion clusters is easier when they follow one another at a sufficient rate that slow ions and electrons can accumulate in the ion duct. Favorable ion behavior occurs when the ions themselves move slowly to the wall and their own space charge confines the electrons until they are captured when the following cluster of fast ions arrives. The compensation of positive ion clusters can be improved by placing electron emitters along their path. The two observed regimes differ in the range of modulation frequencies that cause cluster formation. At low frequencies ($f < 0.5$ MHz) the clusters are decompensated regardless of the gas pressure and the potential of the clusters falls to the minimum only above some pressure level. At high frequencies ($f > 4$ MHz) the potential of the clusters is a minimum over the entire range of variation of the gas pressure.

It is important to note that under suitable conditions electron capture leads to compensation of the current as well as of the space charge of ion clusters.

If, for comparison, we consider a unit cluster of negative ions that is compensated by heavy positive ions rather than by electrons, then the minimum potential for this cluster is considerably higher. The analogous condition for capture $e\varphi_{min} \geq M_+v_b^2/2$ implies that positive ions can be captured only by a very deep potential well, so that the cluster spreads out before it is compensated. Because of the inefficiency of gas compensation for negative ion clusters, a similar cluster of positive ions propagating at the same velocity as the negative ions and coincident with them may be used for this purpose.

In schemes for realizing ion inertial thermonuclear fusion it is proposed, in particular, that phase focusing of the ions be used. The effect of incomplete compensation of clusters on the phase focusing process has been examined [36]. The minimum cluster length may be finite even in the absence of spreading in the ion energies.

2.8 Plasma-Optical Lenses and Transport of Compensated Ion Beams

In many cases focusing of intense compensated ion beams by ordinary electrostatic or magnetic lenses is ineffective or entirely impossible. The difficulty in using an electrostatic lens for this purpose is that the electric field is shielded by the charges in the ion-beam plasma and it becomes impossible to maintain a large potential difference in the gap of the lens. In themselves, magnetic lenses are not very effective since their focal length is proportional to the mass-to-charge ratio of the particles and strong magnetic fields are needed to control the trajectories of heavy ions. The lenses based on the concepts of plasma optics (primarily the principle that the magnetic field lines in a plasma are equipotentials) developed by Morozov [37] are of particular interest in this regard.

The principle of plasma optics. The existence of (macroscopic) electric fields in plasmas is inhibited primarily by the high mobility of the plasma electrons. Favorable conditions for maintenance of an electric field in a plasma occur

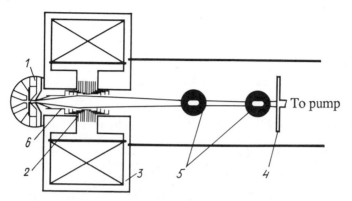

Fig. 2.8. Focusing of an ion beam from a duoplasmatron by a plasma lens: (1) duoplasmatron with extractor; (2) electrodes of the plasma lens; (3) electromagnet; (4) collector; (5) photographic image of the elements of the focused ion beam; (6) compensator – electron source.

when a sufficiently strong magnetic field penetrates the plasma, so that the Larmor radius of the plasma electrons, $\rho_e = mv_ec/(eH)$, is considerably smaller than the characteristic size of the plasma. Then the electrons undergo free movement only along the lines of force of the magnetic field. If a line of force is incident on an electrode with a given potential, then because of the high mobility of the electrons the potential along the entire line is roughly constant and, to within a factor proportional to the electron temperature, is equal to the potential of this electrode. A high electric field may be maintained perpendicular to the magnetic field, however, between lines of force which end up on electrodes with different potentials. These qualitative considerations are combined with the equation of plasma optics

$$\mathbf{E} = -(1/c)\mathbf{u}\mathbf{x}\mathbf{H}, \tag{2.49}$$

which means that a strong electric field can exist in a "cold" plasma ($T_e = 0$) only in the presence of a transverse magnetic field and an electron drift [38]. The standard terminology seems somewhat inadequate in view of the complicated process by which space-charge lenses have been developed [38]. Without minimizing the significance of the stages in their development, in particular the well-known work of D. Gabor, it appears that the action of modern focusing devices should be analyzed in terms of the above considerations from plasma optics.

Plasma lenses are made as follows: current-carrying coils or permanent magnets are used to create the required magnetic field configuration. The field lines are incident on electrodes to which an optimal potential distribution is applied. Electrons from a special compensator are introduced into the lens. The compensator may, for example, be a metal cylinder on whose inner surface part of the beam ions fall and cause ion–electron emission. When a sufficient flux of electrons enters the lens, the magnetic field lines simultaneously become approximate equipotentials. Short lenses of this type have been used, in particular, for focusing beams produced in the accelerator system of a duoplasmatron [39] (see Fig. 2.8). Experiments have been performed with an analogous long lens in which the compensator was a thermionic cathode [39] located inside the lens itself.

Plasma lenses have interesting properties. While ordinary electrostatic lenses, which accelerate or decelerate particles of a particular sign, are focusing lenses, plasma lenses may be focusing or diverging lenses, depending on the direction of the electric field between the axial and peripheral regions. The optical power of a plasma lens is considerably greater than that of a lens with the same magnetic field but without the electron space charge. The focal length of plasma lenses, as of electrostatic lenses, is independent of the mass-to-charge ratio of the particles being focused. A plasma lens can also operate in a self-consistent regime, i.e., such that the required potential difference across the electrodes is created by the ion beam itself, rather than by an external source, as part of the beam strikes these electrodes.

Although these lenses are referred to as plasma lenses and can actually be used when they are filled with plasma, their successful operation only requires an electron component. Positive ions, on the other hand, weaken the field in a lens so a high vacuum must be maintained in it. Then the electrons enter the lens primarily from compensators and not from an ionized gas. Otherwise, slow positive ions are formed along with the electrons. The electric field and, therefore, the optical power of the lens are greatest when a dense electron cloud exists along with the beam ions and the space charge of the beam is overcompensated ($n_e/n_+ \gg 1$).

Plasma lenses can also operate in a pulsed regime. Under these conditions the necessary electron space charge accumulates over a finite time which depends, naturally, on the strength of the compensator. When the gas pressure is excessive, slow ions accumulate in the lens because of ionization of the gas and inhibit its efficient operation.

Complicated oscillatory processes in a plasma lens undoubtedly affect its operation and are as yet little studied. It is also important to reduce aberrations in these lenses to a minimum.

The lens used for focusing and accelerating a 10-kA ion beam with a current density of 50 A/cm^2 in a linear accelerator by Humphreys [40] is an example of the application of the principles of plasma optics. The ion beam propagated in an annular drift tube with accelerating gaps. Figure 2.9 shows the symmetric half of the longitudinal cross section of the tube near one gap. A system of coils creates a magnetic field that is perpendicular to the gap. Part of the beam ions strike the wall and release electrons which oscillate in space and compensate the space charge of the beam. The magnetic field serves to enhance the electron density in the gap, where it is intended to localize a potential difference of 1-2 MV. The field lines which intersect the drift tube near the gap pass through a region where the space charge of the compensating electrons is concentrated and are, therefore, equipotentials. The shape of these field lines is such that the ions are focused both by the magnetic field that pushes the beam away from the coils and by the electric field in the gap. The possibility of focusing a high-current ion beam with a dipole magnetic lens has been demonstrated experimentally.

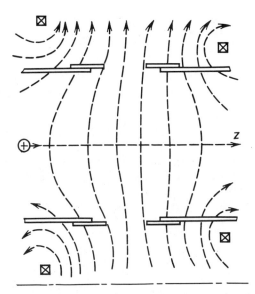

Fig. 2.9. The magnetic field in an accelerating gap of a high-current linear accelerator.

Transport of high-current ion beams. Intense ion beams cannot be transported without compensation of their space charge. This is especially true of high-current ion beams, which must be compensated as soon as they leave the source. In reflex triodes the ion beam captures electrons in the region of the virtual cathode. In magnetically confined diodes the transverse magnetic field prevents electron capture of this sort and electrons must be introduced by other means. One such device is a diode whose negative electrode adjoins a region with enhanced gas pressure. Already in the diode region (pinch-diode) the high-current ion beam is contracted by its own magnetic field so that the ions leave the diode with substantial radial velocities that are proportional to their distances from the axis. Then in the high-pressure gas region the space charge of the ions is compensated, and the ions propagate along rectilinear trajectories and are collected at a certain distance from the diode in a more or less distinct focus. Another possibility is to place a plasma channel, created in advance by a high-power pulsed discharge, in the beam path. The propagation of a pulsed high-current beam along a channel with a finite inductance produces an electric field which slows down the ions. Hence, space charge compensation of these beams must be supplemented by current compensation. The azimuthal field of the discharge current that forms the plasma channel also facilitates transport of the ion beam.

In a number of other devices the electrodes of the source (for example, a diode) are shaped so that focusing proceeds along the natural rectilinear trajectories of the particles (under compensation conditions). A valid discussion of the prospects for focusing compensated high-current beams requires more detailed information on their emittance and residual radial field than has been available up to now. It is often assumed that the temperature of high-current beams is the temperature of the plasma that generates them and is at most a few or a few tens of electron volts. Meanwhile, it is difficult to imagine that the effective temperature of an ion beam produced, say, in the neighborhood of a structure as complicated as a virtual cathode is really so low.

Electron "cooling" of ion beams. The problem of constricting ion beams is related to cooling them, i.e., to reducing their phase volume. Budker [41] proposed a method for cooling a "hot" ion beam by combining it with a "cold" electron beam. The energy dissipation needed to reduce the phase volume takes place through Coulomb collisions between the ions and electrons. This kind of electron cooling, of which several variants can be conceived, is of fundamental importance. In this connection we note the work of Winterberger [42], who has examined a high-current beam of multiply charged ions entering a magnetic field region with gradually converging lines of force. A compensated ion–electron beam whose species are coupled by the action of its space charge is a unified plasma formation. As it enters a magnetic mirror region this plasma formation is compressed. (An azimuthal current develops in the plasma and interacts with the longitudinal component of the magnetic field to produce a radial focusing force.) It was suggested that radiation from multiply charged ions caused by their interaction with the electrons leads to cooling of the electrons and, consequently, to a reduction in the ion temperature which promotes compression of the entire system.

Chapter 3

HIGH-CURRENT ION SOURCES
AND FAST ATOM INJECTORS
FOR THERMONUCLEAR EXPERIMENTS

3.1. General Information

The development of methods for producing powerful fluxes of charged and neutral particles has been stimulated primarily by work in the controlled thermonuclear fusion program. High-current beams of ions and atoms of hydrogen (deuterium) are used to create and heat plasmas to thermonuclear temperatures in magnetic confinement experiments. The development of fast particle injectors for thermonuclear experiments began at the end of the 1950s, when construction of the classical magnetic mirrors began ("Ogra" in the USSR and DCX in the USA). In the 1970s the need to build means for non-Ohmic heating of plasmas in tokamaks was the stimulus for great progress in the development of injectors. Whereas the power of injection systems in the first tokamaks was tens of kilowatts, today these systems reach several megawatts [43], and in the largest tokamaks of the 1980s (TFTR, JET, JT-60, T-15, Tore Supra), tens of megawatts. Designs for demonstration thermonuclear reactors (INTOR, OTR, MFTF-B) include neutral beam injection systems with powers of up to 100 MW (see Table 3.1).

The injectors in thermonuclear experiments can perform a number of tasks, including:

1) heating the plasma to thermonuclear temperatures in closed magnetic systems (tokamaks and stellarators) and creating high-temperature plasmas in open magnetic systems (classical and ambipolar mirrors);

2) supplying "fuel" to a thermonuclear experiment;

3) controlling the rate of thermonuclear reactions; and

4) producing a noninductive current in the toroidal plasma of a tokamak.

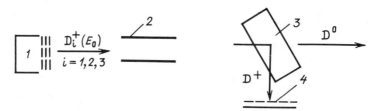

Fig. 3.1. A schematic illustration of an injector based on positive ions:
(1) ion source (gaseous discharge chamber and ion-optical system); (2) neutralizer; (3) deflecting magnet; (4) energy recovery system.

TABLE 3.1. Neutral Beam Injectors in Thermonuclear Experiments

Experiment and country	P_0, MW	Particle energy E_0, keV	No. of injectors	No. of ion sources per injector	Pulse duration t, sec
Achieved parameters (1976-1984)					
ORMAK (USA)	0.36	30	1	1	0.05
TFR (France)	0.45	34	2	1	0.05
T-11 (USSR)	1.2	30	2	1	0.03
PLT (USA)	2.5	35	4	1	0.15
ASDEX (FRG)	3.0	40	2	2	0.2
PDX (USA)	7.0	50	4	2	0.15
D-III (USA)	7.0	70	4	2	0.25
Planned parameters (1985-1990)					
TETR (USA)	32	120	4–6	3	1.5
JET	10	80	2	8	10
JT-70 (Japan)	20	75	14	2	10
T-15 (USSR)	10	80	3	2	1.5
TORE SUPRA (France)	4	120	2	2	5
MFTF-B (USA)	60	75	–	–	10
INTOR (IAEA)	75	175	4	4	10

Injectors can be used in thermonuclear experiments with different goals, i.e., in research devices, demonstration devices, or reactors. The basic parameters and specifications of injection systems depend to a great extent on the purpose of the experiment. The power P_0 of the particle flux delivered to the plasma lies in the range 10-100 MW (see Fig. 3.1). It is determined by the size of the machine, the plasma volume, and the energy confinement time of the particles in the system. The energy E_0 of the injected particles lies in the range 40 keV-1 MeV. It depends on the type and size of the machine. For tokamaks it is up to 200 keV and for open mirrors, up to 1 MeV. E_0 is determined primarily by the penetration depth of the particles in the plasma and by the distribution of the power density deposited by the injected beam along its path. The duration of the injection pulse t_i varies from tenths of a second to hundreds of seconds. It is determined by the purpose of the machine.

The power efficiency of an injection system (the ratio of the power of the flux delivered to the plasma to the total power expended by the injector, $\eta_{inj} = P_0/P_{in}$) is of great importance in choosing heating methods. It will be decisive in reactors with continuous injection, where it must be at least 0.7. For ignition reactors the efficiency is not so decisive, since injectors will operate at full power for only a small fraction ($\approx 5\%$) of the working pulse. The efficiency η_{inj} depends mainly on the particle energy E_0. Systems for recovering the energy of ions not used for injection are being studied in order to raise the efficiency.

In addition to the above parameters, a decisive role in creating and heating plasmas is played by the energy spectrum of the beam of particles introduced into the plasma and by the fluxes of heavy particle impurities and background gas. The energy spectrum of the injected particles affects their penetration depth into the plasma and the power density profile. It depends on the species mix of the primary ion beam extracted from the ion source (the relative amounts of H_1^+, H_2^+, and H_3^+). At this time sources have been developed which produce beams with a 90% proton content.

The presence of heavy element impurities in an injected beam may affect the Z_{eff} of the plasma. According to the latest data, the amount of light element impurities (oxygen, carbon, etc.) is 0.2-0.3%, while that of heavy elements (copper, zinc, molybdenum, tungsten) is less than 0.01%. The flux of background gas from an injector is extremely critical for open mirrors (where it must be less than 10^{-3} of the fast particle flux) and less so for tokamaks (where it can be up to 10%). It places certain demands on the geometry of the injection path and on the methods for vacuum pumping. The length of the injection path depends to a great extent on the type of machine, the dimensions of its magnetic system, and its purpose (the presence of blankets, shielding, etc.). The transverse cross section of an injected beam is determined by the dimensions of the entry ports in the magnetic system and shielding (for a reactor). The power density of the incident flux is as high as 2-4 kW/cm^2, so that the area of the injection ports need not exceed 1-2% of the total area of the vacuum vessel walls. These parameters are the basis for injection system design.

An injection system for a thermonuclear experiment is composed of a number of simultaneously operating, autonomous injectors that together provide the required power P_0. Each injector has an independent vacuum system and contains several modular ion sources with autonomous electrical power and breakdown protection systems. An automatic control system ensures that all the ion sources operate synchronously. This injection system design makes it possible to choose the optimal dimensions for the injectors and ion sources, to match the electrical power needs of the injector components to commercially available equipment, and to repair and replace failed parts without curtailing operation of the entire system.

The demand for maximum power efficiency of an injection system over the range of atom energies needed in fusion devices (from 40 keV to 1 MeV) leads to injection systems that work in two ways: production of atoms by charge exchange of positive ions or by conversion of negative ions. Both methods are fairly effective for producing neutral atom beams in their respective energy ranges: 20-200 keV D^0 and 200 keV or above, respectively.

In an injector based on positive ions (Fig. 3.1) an ion source (or group of sources) creates and shapes a primary ion beam with a nominal energy E_0. On passing through a neutralizer (charge exchange target) a certain fraction of the ions undergoes charge exchange. The resulting flux of fast neutrals is aimed at the thermonuclear plasma, and the remaining ions are deflected by a magnetic field. The power efficiency of a positive ion system is given by

$$\frac{P_0}{P_{inj} - P_{recov}} = \frac{\overset{0.95}{\eta_{source}} \overset{0.90}{\eta_{duct^+}} \eta_{neut} \overset{0.97}{\eta_{reion}} \overset{0.90}{\eta_{duct^0}}}{1 - \eta_{source}\, \eta_{duct^+} (1 - \eta_{neut})\, \underset{0.85}{\eta_{recov}}},$$

where P_0 is the neutral beam power delivered to the plasma, P_{inj} is the beam power delivered to the injector, P_{recov} is the energy expended in the ion recovery system, η_{source} is the ion source efficiency, η_{duct^+} is the efficiency of the positive ion beam duct, η_{duct^0} is the efficiency of the neutral beam (fast atom) duct, η_{reion} is the efficiency relative to reionization (the loss of fast atoms by ionization), η_{neut} is the neutralizer ($D^+ \rightarrow D^0$ charge exchange target) efficiency, and η_{recov} is the efficiency of the ion recovery system.

In an injector based on negative ions (Fig. 3.2) the source creates and shapes a beam of negative ions with a low energy E_1. In the final accelerator system the negative ions acquire the nominal energy E_0. Then the negative ions are stripped (an electron is removed) in a special target (gaseous, metal vapor, plasma, or laser) and the resulting flux of atoms with energies E_0 is delivered to the thermonuclear plasma. The remaining small number of ions is deflected by a magnetic field. The power efficiency of a negative-ion-based injector is defined as

$$\frac{P_0}{P_{inj} - P_{recov}} = \frac{\eta_{source}\, \eta_{duct^+} \eta_{cx}\, \eta_{duct^-} \eta_{accel}\, \eta_{str}\, \eta_{duct^-}}{1 - \eta_{source}\, \eta_{duct^+} \eta_{cx}\, \eta_{duct^-} \eta_{accel}\, \eta_{duct^-} (1 - \eta_{str})\, \eta_{recov}},$$

where η_{cx} is the efficiency of the charge exchange target ($D^+ \rightarrow D^-$), η_{duct^-} is the efficiency of the negative ion beam duct, η_{accel} is the efficiency of the system for acceleration from E_1 to E_0, and η_{str} is the efficiency of the stripping cell ($D^- \rightarrow D^0$).

On the whole, positive-ion-based injectors are relatively simple and the technology for them is well developed. Since the energy of the particles is presently limited to 120-160 keV, all existing injection systems on modern machines use charge exchange. The efficiency of an injector, however, drops rapidly as the particle energy is raised (Fig. 3.3). At high energies (above 200 keV), therefore, injectors based on negative ions become energetically favorable. They are still in the research stage, but if injection of high-energy atoms is required in large machines negative-ion-based accelerators will find application, especially for generating noninductive currents in tokamaks. Naturally, the technology of these sources is more complicated and expensive.

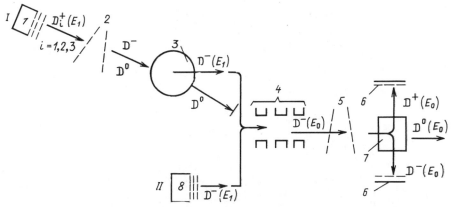

Fig. 3.2. The layout of an injector based on negative ions: (I) D^- ions are obtained by double charge exchange of positive ions on a metal vapor target $D^+ \to D^-$; (II) D^- ions are extracted directly from the ion source plasma. (1) D^+ source (gaseous discharge vessel and ion-optical system); (2) charge exchange target; (3) magnet I; (4) final accelerator system; (5) stripping cell; (6) ion recovery system; (7) deflecting magnet II; (8) D^- source.

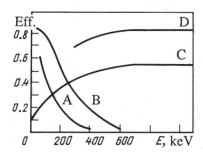

Fig. 3.3. The dependence of the injector efficiency on the ion energy E_0 for deuterium: (A) positive-ion-based injector; (B) the injector of A with ion energy recovery (at an efficiency $\eta_{recov} = 85\%$); (C) negative-ion-based injector with a gas or metal vapor $D^- \to D^0$ stripping target; (D) injector with a laser target for $D^- \to D^0$ stripping.

3.2. Positive-Ion-Based Injectors

The successful operation of injection experiments (PLT, T-11, ISX-B, D-III, TFTR, JET, cf. Table 3.1) and the tendency to develop injectors with higher unit powers, increase the particle energies (up to 200 keV), and reach quasistationary operating regimes have led to something of a unified injector design, which has been developed in a number of countries with relatively small structural differences. A conceptual diagram of a typical injector is shown in Fig. 3.4.

The components along the path followed by the ions and atoms are located in a large rectangular or cylindrical vacuum vessel. As a rule, several ion sources are mounted at the end of this vessel. In some designs (T-15, JT-60, D-III) the ion

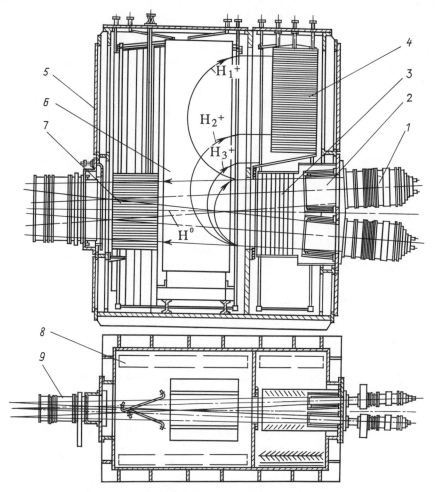

Fig. 3.4. A diagram of an injector: (1) ion source; (2) neutralizer channel; (3) louvered magnetic shield; (4) collector for deflected ions; (5) vacuum vessel; (6) deflecting magnet; (7) movable atom collector; (8) cryopanel pumps; (9) neutral beam duct.

sources are mounted in a vertical array along the long axis of the rectangular beam inlet ports on the tokamak. In other designs (TFTR, ZEPHYR) the ion sources are located along the horizontal axis of the input window. Combinations are possible, as in the JET injectors, which have four sources along the vertical in two rows. The positive ions undergo charge exchange in the neutralizer chamber, which usually is immediately adjacent to the ion source. An equilibrium charge exchange target is formed from the gas flowing out of the ion source since the gas efficiency of the ion source is about 50%. The neutralizer walls serve simultaneously as magnetic shields. At the outlet of the neutralizer there is a louvered magnetic shield along whose length the gas is pumped out. The beam atoms and unneutralized ions are separated by an electromagnet. The deflected ions fall on current collectors, and the fast atoms pass through the beam duct into the tokamak vessel. In order to ensure efficient transport of the fast atom beam along the injector duct, the influxes of gas from the ion source through the neutralizer channel and from the ion current collectors are pumped by a differential pumping system. The injector vessel is divided by thin-walled barriers into several vacuum sectors each of which contains cryogenic pumping assemblies. The overall pumping speed reaches several million liters per second. The injectors on TFTR, JET, T-15 and the IREK test stand are typical representatives of this design.

3.3. Positive Ion Sources

The considerable progress in developing high-current ion sources over the past few years has meant that the specifications for the injection systems on thermonuclear experiments could be met.

An ion source consists of two functional parts: a gaseous discharge chamber in which a low-voltage, low-pressure diffuse discharge creates a plasma ion emitter and an electrostatic ion-optical system which extracts ions from the plasma, accelerates them, and shapes a directed beam.

High-current ion beam production in the new generation of ion sources (duopigatrons, periplasmatrons, sources without an external magnetic field, sources with an edge magnetic field) is based on (a) gaseous discharge plasma emitters with a large surface area for emission of ions (hundreds of square centimeters) over which the plasma density is kept highly uniform (better than $\pm 5\%$) to provide ions with a current density $j_i \approx 0.5$ A/cm^2, and (b) multiaperture (hundreds and thousands of apertures), three- or four-electrode ion-optical systems which form beams with large transverse cross sections and low divergence angles (less than 1°). The sources now in use and under development yield hydrogen ion beams with currents above 100 A, particle energies of 10-120 (150) keV, and pulse durations of from tens of milliseconds to several seconds.

In thermonuclear experiments an important characteristic of a plasma emitter is the species mix of the extracted ion flux, i.e., the relative amounts of H_1^+, H_2^+, and H_3^+ ions. An H_1^+ fraction of greater than 80% is desirable. The conditions under which an ion source in an injector assembly operates impose certain demands on its gas and energy efficiency. The gas efficiency of the discharge emitter must be at least 50% for the typical pressure of below 1.3 Pa in gaseous discharge vessels. The energy efficiency of the discharge depends on the type of ion source and may reach 6-8 A/kW [46].

High-current sources have been developed with and without a magnetic field. The use of a magnetic field usually ensures better confinement of the ions and fast cathode electrons in the volume of the gaseous discharge vessel and thereby yields a higher energy and gas efficiency for the discharge. However, a magnetic field does not produce plasma emitters with good uniformity over a large area and causes a significant level of oscillations in the plasma at frequencies of 10^4-10^6 Hz. In sources without an external magnetic field, on the other hand, it is possible to obtain uniform emitters with areas of hundreds of square centimeters with almost no

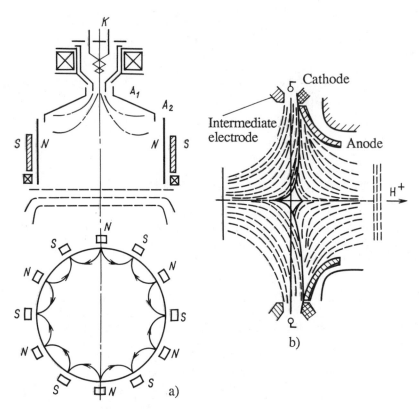

Fig. 3.5. The design of the gaseous discharge vessel and the magnetic field configuration in a duopigatron (a) and in a periplasmatron (b).

modulation of the ion current, although with a lower discharge efficiency [47]. In recent years sources with multipole edge magnetic fields (magnetic walls) have been developed in which the magnetic field is localized near the walls of the gaseous discharge chamber and is absent in the central region. In these sources charged-particle losses from the discharge plasma are reduced (so the discharge efficiency is higher) and good plasma uniformity is maintained over the emission boundary [48].

The above types of ion sources (duopigatron, periplasmatron, sources without external fields and with an edge field) differ primarily in the organization of the gaseous discharge – the ion emitter – while their ion-optical systems are the same. We now briefly examine the basic construction of each type of source and analyze the operation of sources with an edge magnetic field in more detail.

Duopigatron sources. The modern duopigatron is an extension of the earlier duoplasmatrons (see, for example, [1]) in which the degree of ionization is enhanced by subjecting the plasma column to electrostatic and magnetic compression with the aid of diaphragms and a nonuniform magnetic field which increases toward a small-diameter anode aperture. The duoplasmatron produced a dense plasma which made it possible to obtain an ion emission current of tens of amperes per cm^2, i.e., it formed a point ion source with little flow of cold gas from the gaseous discharge chamber into the accelerator gap. However, single aperture ion-optical systems were unable to form beams with such a high current density, and it was necessary to install a special plasma expander beyond the anode aperture in order to reduce the current density j_+ (see Chapter 2).

For producing high-current beams the duoplasmatron was then modified so as to create a plasma emitter with a large area and reduced current density j_+ (about 0.5 A/cm^2). The fundamental improvement was the installation beyond the anode of an additional large-diameter vessel at the anode potential and an anticathode, which made it possible to obtain a discharge regime with a magnetic field and oscillating electrons (i.e., a PIG discharge). The new model (Fig. 3.5a) was called the duopigatron. The extraction electrode of the multiaperture ion-optical system located at the end of the gaseous discharge chamber serves as an anticathode.

The plasma generator consists of a heated cathode, an intermediate electrode, two anodes, and an anticathode (which is the first, extraction or emitting, electrode of the ion-optical system). The intermediate electrode with a specially shaped channel serves simultaneously as a pole of an electromagnet that creates a nonuniform longitudinal magnetic field in the anode region which falls off toward the anticathode. In the heated cathode region the magnetic field is practically absent.

The discharge voltage is applied between the anodes and heated cathode. The intermediate electrode and anticathode are connected to the discharge rectifier circuit through suitable resistors and are at the floating potential. Primary electrons emitted from the cathode are accelerated toward the intermediate electrode, acquire a relatively low energy (below 30 eV), and form the cathode plasma. The cathode plasma is separated from the plasma in the anode region by a double layer that forms in the intermediate electrode region (Fig. 3.5a).

The plasma density inside the intermediate electrode hollow is determined by the discharge current and gas pressure. Electrons from the cathode plasma pass through the intermediate electrode channel and are accelerated in the anode region, thereby acquiring a fairly high energy. Their motion in the anode region is limited by the magnetic field, and they oscillate between the intermediate electrode and the anticathode, thus forming a dense plasma. The gaseous discharge plasma of a duopigatron, therefore, consists of two parts, a cathode plasma and an anode plasma, separated by a double layer region. The cathode plasma serves as the source of ionizing electrons for the anode plasma. The latter, in turn, emits the ions in the beam formed by the ion-optical system.

A longitudinal magnetic field makes it possible to organize a PIG discharge. The improved charged particle confinement in this type of configuration compared to an ion source without an external magnetic field makes it possible to reduce the pressure in the anode region to 0.7 Pa and raise the gas efficiency to 50%. The longitudinal magnetic field causes nonuniformities in the radial distribution of j_+ and the appearance of a radial electric field in the plasma. The rotation of the plasma column in the crossed E and H fields produces an instability that leads to a modulation of j_+ that is especially strong at the edge of the column. It is, therefore, natural to try to make the longitudinal magnetic field as small as possible in the anode plasma region. Combining the longitudinal magnetic field with a transverse multipole magnetic field on the edge of the anode region (Fig. 3.5a) reduces the ion and electron losses from the plasma and considerably improves the uniformity of the j_+ distribution on the emitting surface [49]. The transverse magnetic field is produced by a set (from 12 to 32) of permanent magnets with alternating poles, symmetrically positioned along the generatrix of the cylindrical anode. The field is about 0.1 T on the anode surface and falls off rapidly in a radial direction toward the discharge vessel center.

Duopigatrons have been built with ion-optical system diameters of up to 28 cm, ion beam currents of up to 100 A, ion energies of 30-50 keV, fractions of atomic ions of up to 85%, gas efficiencies of up to 60%, and energy efficiencies of about 1.1 A/kW [50].

Periplasmatron. Periplasmatron high-current ion sources were developed at Fontenay-aux-Roses (France). The basic ideas behind the construction of this type of source are the following:

1) the creation of a uniform plasma in the ion emission region by delivering ionizing electrons to the plasma from the outer radial boundary, and

2) confining these electrons with the aid of magnetic shielding for the anodes, which are located in a cusp magnetic field.

Cylindrical and rectangular sources have been built. The cusp magnetic field (Fig. 3.5b) is created by two coils with currents flowing in opposite directions. The magnetic field is zero in the center of the discharge vessel.

A cathode in the form of heated filaments is placed on the periphery around the discharge chamber in the hollow of an intermediate electrode that is made of soft steel and serves simultaneously as a shield against the magnetic field. A configuration resembling an antimirror magnetic field results [52]. Primary electrons are introduced into the discharge along the median plane of the cusp field. The extraction electrode of the ion-optical system and the electrode on the opposite side serve as electrostatic electron reflectors and electrons must diffuse across the magnetic field in order to reach the anode. The magnetic field strength near the anodes is about $4 \cdot 10^{-3}$ T. Hydrogen enters the discharge chamber directly. The cathode consists of several filaments, wound over the exterior and with independent heater currents, each of which is connected through a resistance of about 1 Ω to the discharge rectifier.

The periplasmatron has a number of advantages:

1) the edge location of the cathode makes it possible to use a large number of cathode filaments;

2) the weak cusp magnetic field ensures confinement of the ionizing electrons without noticeably affecting the uniformity and stability of the plasma;

3) the direct thermal emission from the heated cathode filaments, which might cause a large heat load, does not strike the extraction electrode of the ion-optical system; and

4) the cathode filaments and insulators are well protected from flows of reverse electrons from the accelerating gap.

Experiments have shown that the plasma uniformity depends on the location of the null point of the magnetic field surface. The plasma is more uniform when this point lies in the median plane of the source or closer to the surface of the extraction electrode. When the null point is far from the emitting surface, the electrons concentrate mainly on the edge and the plasma density at the center falls.

The operating pressure of hydrogen in the vessel is 1.3 Pa. A discharge is ignited with or without a magnetic field. When a field is present, the ion production efficiency is considerably higher. It is important that the anode surface coincide with a magnetic line of force. Then the plasma density is proportional to the discharge current.

The first periplasmatrons were cylindrical with a plasma diameter of 20 cm [53]. The three-electrode ion-optical system had 450 apertures, a diameter of 14.5 cm, and an emitting area of 107 cm^2. A 20-A beam at an energy of 30 keV was extracted from a discharge with a current of about 250 A and $U_{disch} = 90$ V. The discharge efficiency was as high as 1.1 A/kW.

A later model of the periplasmatron is rectangular in shape [53] and produces a uniform plasma emitter with $j_+ \approx 0.22$ A/cm^2 over an area of 12×38 cm^2 from $I_{disch} = 700$ A and $U_{disch} = 110$ V with a pulse duration of 1 s [53].

Ion sources without an external magnetic field and with an edge magnetic field. The first ion sources without an external magnetic field appeared in 1972. They drew immediate attention because of certain advantages, namely, the relatively simplicity of making a plasma emitter with a large surface area and the absence of a magnetic field [47].

The design of the source is fairly simple (Fig. 3.6). The gas discharge chamber (unit I) includes the cathode assembly, anode flange, walls, and gas feed system.

The emitting plasma is produced by a low-pressure diffuse discharge that is ignited in the discharge vessel with a distributed cathode without applying an external field. The end and side walls of the gaseous discharge vessel are electrically isolated from one another. The front end wall is the extraction (emitting) electrode of the ion-optical system. Cathodes made of 0.5-0.7-mm-diameter tungsten wire in the shape of hairpins are mounted around the inside edge of the vessel. The total area of this cathode may be tens of square centimeters. In different designs the

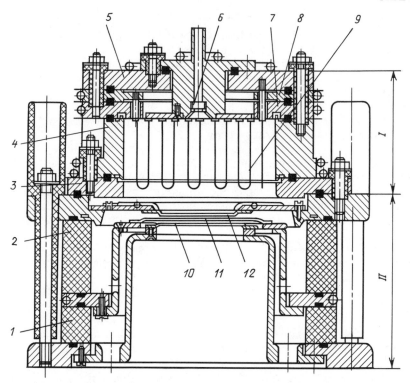

Fig. 3.6. An ion source without an external magnetic field (I is the gaseous discharge vessel, II is the ion-optical system): (1),(2) insulators; (3) anode flange; (4) housing; (5),(8) cathode flanges; (6) gas feed; (7) shield; (9) cathode (pins); (10)-(12) electrodes of the ion-optical system.

plane of the hairpins is either along the vessel walls or perpendicular to them. In the latter case the number of hairpins (and therefore the area of the cathode) is greater, and a localized magnetic field of several thousandths of a Tesla develops near the hairpins.

An anode in the form of a rectangular plate is located near the extraction electrode. In some designs the rear end wall of the discharge vessel serves as an anode. This is not of fundamental importance for the discharge, but locating the anode on the rear wall makes it possible to prevent formation of cathode spots which sometimes appear during breakdown of the ion-optical system when the rear wall is insulated. Breakdowns in the ion-optical system frequently cause formation of a strong flux of reverse electrons which can lower the potential of the rear wall enough to cause unipolar arcs. The other parts of the discharge vessel usually are at the floating potential.

Experimental studies of the characteristics of the discharge plasma have revealed the basic discharge behavior. There is special interest in the effect of the discharge parameters (its geometric and plasma characteristics) on the characteristics of the plasma near the emission boundary, since the main problem is to create a large-area, uniform plasma emitter that delivers ions to the emission boundary at a given current density j_+. The plasmas in gaseous discharge vessels of different sizes and configurations (cylindrical with diameters of 10-14 cm, rectangular with cross sections of 12×22, 12×50, and 27×47 cm^2) have been studied.

In order to analyze the experimental data and search for discharge vessel design solutions, a simple model of a uniform hydrogen discharge was developed based on the processes leading to the formation and loss of charged particles [46]. The discharge vessel occupies a volume (V) with a circular or rectangular cross section and contains a heated cathode of area S_K that emits electrons, an anode of area S_A, and walls of area S_W at the floating potential. The cathode electrons are accelerated in a double cathode layer to energies equal to the potential difference between the plasma and cathode. These electrons actually ionize the gas, and the energy required to maintain the electron and ion temperatures (T_e and T_i) is delivered to the discharge through them. In low-pressure discharges ($p_0 = 1.3$ Pa) the condition $T_e > T_i$ is maintained for electron densities of 10^{12}-10^{13} cm^{-3}. When T_e = 5–10 eV, the plasma electrons make a significant contribution to ionization of the gas.

A detailed calculation of the particle and energy balance in the discharge has been made by Kulygin and Panasenkov [54]. The results were used to choose the geometric and plasma parameters of the discharge for an optimum ion emission current density with minimum current, voltage, and gas pressure in the discharge, and these choices were subsequently tested experimentally. The effects of the discharge parameters on the ion species mix and on the energy and gas efficiency of the discharge were established.

The principal energy loss from the plasma in an ion source without an external magnetic field is through poor confinement of charged particles in the discharge vessel volume and their rapid escape to the walls. The important parameters are the following: the ratio of the volume V_{pl} of the plasma in which charged particles are created to the area of all the walls S_L on which ions are lost, $\eta = V_{pl}/S_L$, and the relative area of the cathode, $\kappa = S_K/S_L$. Increasing η causes an increase in the proton fraction in the extracted ion beam and a decrease in the hydrogen density required in the discharge vessel (Fig. 3.7). At the same time, stable operation of the discharge requires that the relative area of the cathode κ exceed some minimum value ($\kappa > 0.04$). The energy efficiency of the discharge for ion formation, which may be defined as $F_+ = I_+/W_d$ (where I_+ is the total ion current produced in the discharge volume and W_d is the discharge power), may reach a peak of 8 A/kW for the optimum $\kappa = 0.03$-0.04. Thus, the cost of an ion in the discharge, $C_+ = 1/F_+$, is about 125 eV.

The discharge efficiency with respect to the ion beam extracted from the source $F_i = I_{H^+}/W_d$ (where I_{H^+} is the beam current) may be considerably lower than F_+ since in a uniform plasma the ratio $F_i{:}F_+ = S_{em}{:}S_L$, where S_{em} is the area of the slots in the extraction electrode of the ion-optical system. In the IBM-5 source (30 A, 30 kV), for example, $S_{em}{:}S_L = 1{:}14$ and $F_i = 0.55$ A/kW. [That is, the cost of an accelerated ion (at the extraction grid) is $C_i = 1/F_i \approx 1.8$ keV.]

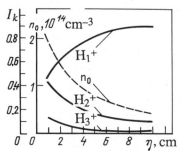

Fig. 3.7. The dependence of the relative amounts (I_k) of H_1^+, H_2^+, and H_3^+ ions in the extracted beam and of the gas density (n_0) in the discharge vessel on $\eta = V_{pl}/S_L$ (calculated for an ion emission current density $j_+ = 0.5$ A/cm^2).

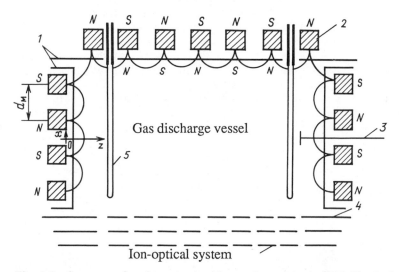

Fig. 3.8. Structure of an ion source with an edge magnetic field ("bucket" source): (1) gaseous discharge vessel wall; (2) permanent magnets; (3) target probe; (4) first (extraction) electrode; (5) cathode.

In order to reduce the energy expenditure in the ion-source discharge and increase the proton yield and gas efficiency it is necessary to improve the charged particle confinement in the discharge vessel by reducing the effective area for ion losses and trying to maximize the ratio S_{em}/S_L. This purpose is served by using a vessel with magnetic walls, near which a fairly high magnetic field prevents loss of charged particles from the plasma volume but falls off rapidly toward the discharge center so that there is no magnetic field in the bulk of the vessel. The idea of using such an edge field was first proposed by Moore [49], extended by MacKenzie [55], and employed in the duopigatron. In order to create an edge field which drops off rapidly toward the discharge vessel center, the vessel is covered with permanent magnets that have alternating poles. Only the emitting (extraction) end of the dis-

charge vessel remains open (Fig. 3.8). The vessel surfaces that are covered by the magnetic field form the anode, and the first (extraction or emitting) electrode of the ion-optical system, is either connected to the cathode or is at the floating potential.

As an example we now consider the characteristics of an ion source with an edge magnetic field, or "bucket" source (Fig. 3.8), whose parameters were determined by the task of the source: to double the ion beam current (to 70 A) while maintaining the discharge power at the same level as in an ion source without an external magnetic field [48, 56].

The discharge vessel is a water-cooled welded stainless steel bucket with a rectangular cross section of 20×33 cm^2 and a height of 12 cm. The cross section of the vessel was chosen on the basis of the requirement that the uniform emission surface of the plasma must be 12×25 cm^2, while the transition regions at the side walls and back end cover, where the magnetic field is higher ($10\text{-}15 \cdot 10^{-4}$ T), correspond to a rapid drop in the plasma density (over a distance of 4 cm). Two rows of cathodes made of 1-mm-diameter tungsten wire with a total emission area of up to 50 cm^2 are located on the back cover along the long sides of the discharge vessel.

The magnetic system for the source is constructed of permanent magnets in the form of parallelepipeds made of magnetically hard barium ferrite with a field near the surface at the pole of about 0.14 T, arranged with alternating poles in a magnetic circuit having a step size $d_M \approx 3$ cm. (The width of a magnet pole is about 8 mm.) In the multipole system the magnetic field falls rapidly with increasing distance from the surface of the magnets. The field $B < 10^{-3}$ T at a distance of $1.5 d_M$ from the wall.

Two variants of the magnetic field geometry have been studied. The configuration in which the magnets are placed parallel to the emission surface has the significant disadvantage that the field from the lower row of magnets is not adequately compensated and in the region of the emitting surface it penetrates deeply toward the axis, narrowing the field-free region (more precisely, the region where $B \leq 10^{-3}$ T). Hence, the variant in which the magnets are mounted on the side walls of the discharge vessel perpendicular to the emitting surface is more effective.

The distribution of the plasma parameters at the emitting surface is usually studied with the aid of a set of plane Langmuir probes mounted on the extraction electrode of the ion-optical system. Several movable cylindrical probes are also mounted in the discharge vessel for measuring the local plasma parameters and on the vessel wall midway between the magnets (isolated from the wall by a large plane target probe) for determining the ion and electron currents to the part of the anode surface that is shielded by the transverse magnetic field.

The discharge characteristics for different hydrogen pressures p_0 in the discharge vessel are shown in Fig. 3.9. For pressures below 0.4 Pa there is a substantial rise in the discharge voltage U_d and a certain saturation in the ion current density to the probe j_+ when the discharge current I_d is raised since the hydrogen density in the vessel is too low. When p_0 is raised, it is possible to move toward

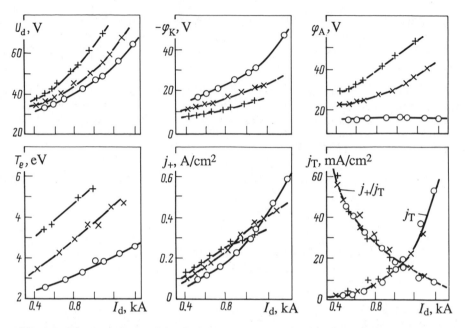

Fig. 3.9. The dependence of the discharge voltage U_d, floating potential of the cathode (φ_K) and anode (φ_A), electron temperature (T_e), and ion current densities to the emitting surface (j_+) and target probe (j_T) on the discharge current for different hydrogen pressures in the discharge vessel: (+) $p_0 = 0.29$ Pa; (×) 0.44 Pa; (o) 0.86 Pa.

larger j_+, and U_d and the electron temperature T_e both decrease. In almost all cases the plasma potential φ_{pl} is close to the anode potential (φ_A), exceeding it by 1-3 V. φ_A is thus close to the potential difference between the plasma and the end of the vessel that floats relative to it ($\varphi_{end} = 0$), which is determined by the condition that the ion and electron currents to the end be equal, i.e., $I_{+,end} + I_{e,end} = 0$. If the ion current $I_{+,end}$ were compensated only by the current of plasma electrons at temperature T_e, then it would be easy to show that φ_A must be near $4T_e/e$. As can be seen from Fig. 3.9, when p_0 is lowered to 0.3 Pa, the ratio $e\varphi_A/T_e$ increases to 6-7. This indicates that at low pressures the fraction of electrons coming directly from the cathode and contributing to the current $I_{e,end}$ increases, so that the cathode potential φ_K approaches the end wall potential.

Measurements of the distribution of j_+ on the emitting surface show that j_+ is uniform with good accuracy (Fig. 3.10) over the 12×25 cm^2 emitting area. A rapid drop in the density begins at a distance of 4-4.5 cm, or about $1.5d_M$, from the wall. The optimum pressure for an ion source with an edge magnetic field is $p_0 \approx 0.8$ Pa (Fig. 3.11). The slight reduction in j_+ with increased pressure at low discharge currents is related to the plasma density gradient in the direction of the end of the vessel since the mean free path of the ions before collisions with neutrals is $\lambda_{in} \approx 0.8/p_0$, or 1-2 cm in this pressure range.

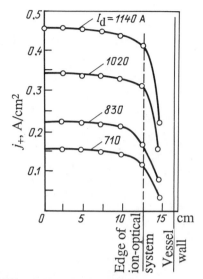

Fig. 3.10. The distribution of the ion current density over the emitting surface along the long side of the discharge vessel for $p_0 = 0.8$ Pa.

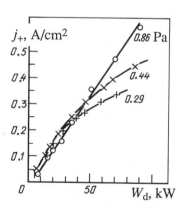

Fig. 3.11. The dependence of j_+ on the discharge power for different pressures in the discharge vessel.

The current density ratio j_+/j_T falls noticeably when I_d is raised. This ratio is practically independent of the hydrogen pressure in the vessel, so that j_T depends on j_+ quadratically. This fact indicates that the ion diffusion coefficient in a multipole magnetic field depends on the plasma density.

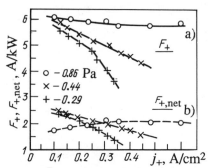

Fig. 3.12. The dependence of the discharge efficiency for total ion production F_+ (a) and "net" efficiency $F_{+,\text{net}}$ (b) on the emission ion current density for different pressures in the discharge vessel.

The plasma parameters at different distances from the wall in the region where the field component $B_x(z)$ varies from $6 \cdot 10^{-2}$ to $5 \cdot 10^{-4}$ T have been measured using a target probe that moves in the midplane between the magnets. A sharp drop in the ion saturation current sets in at $z < 3.5$ cm ($B_x \geq 3 \cdot 10^{-3}$ T). T_e also decreases in this region. The probe characteristics indicate that there is a significant number of fast cathode electrons in the region $z > 3.5$ cm and that they are absent in the region $z < 3$ cm, where $B_x > 5 \cdot 10^{-3}$ T. A fast electron incident on the magnetic field will be reflected at some distance from the wall and, if its time of flight in the field is much less than the time between elastic collisions that cause it to be scattered, then, for some distance z_0 there should be no fast electrons in the region $z < z_0$. Calculations show that $z_0 \approx 3$ cm.

Knowing the ion saturation current in the plasma, the distribution of j_+ over the emitting (extraction) end of the vessel, and j_T, it is possible to estimate the total ion current created in the discharge vessel volume and lost at the end, cathode, and body of the vessel at the anode potential ($\varphi_{\text{pl}} - \varphi_A > 0$): $I_+ = I_{+,\text{end}} + I_{+K} + I_{+A}$, where the area of the cathodes and holders is about 120 cm² and the area of the vessel body is 2000 cm². The ratio of I_+ to the discharge power W_d gives the efficiency of the discharge for producing ions (F_+) and the ratio of the ion current onto the part of the emitting (extraction) end where j_+ is uniform to W_d gives the "net" or useful efficiency ($F_{+,\text{net}}$) (see Fig. 3.12). For $p_0 = 0.86$ Pa, F_+ exceeds 6 A/kW, which is close to the calculated limit. At lower pressures, F_+ decreases with increasing j_+. $F_{+,\text{net}}$ reaches 2-2.5 A/kW and is roughly twice the value of $F_{+,\text{net}}$ obtained from sources without an external field. A slight reduction in $F_{+,\text{net}}$ with increasing p_0 is caused by an increased plasma density gradient.

The values of I_+ and j_+ can be used to estimate the effective area for ion losses, which is about 750 cm². Thus, η and κ for the source with an external field ('bucket' source) are given by $\eta = 3.5$ cm and $\kappa \cong 0.06$. The ion losses to the vessel walls in a source with an external field are still rather large and are mainly related to the fact that the plasma potential is positive relative to the anode. In fact, in this type of system the ions moving across the field to the walls are weakly magnetized, since at the characteristic vessel pressure of 0.665 Pa, $\omega_{ci}\tau_{in} \cong 5 \cdot 10^{-7}$ V, where ω_{ci} is the ion cyclotron frequency, τ_{in} is the ion–neutral collision time, and the perpendicular component B_x near the walls only reaches 0.06 T. Thus, an effective method for reducing ion losses from the plasma to the anode is to make the plasma potential lower than the anode potential.

We now consider the balance of the electron currents to the cathode and anode [54]. From the theory of the double layer (see [1], for example) it follows that when the cathode operates in the space-charge regime, the electron emission current density is determined by the current density of the ions going to the cathode from the plasma (the Langmuir criterion):

$$j_{e\text{K}} \approx j_+ \sqrt{M_+/m_e}.$$

Since $j_+ = 0.4en_e\sqrt{2T_e/M_+}$, the electron current from the cathode is

$$I_{e\text{K}} \approx 0.4\,en_e\sqrt{2T_e/m_e}\;S_\text{K}.$$

When $\Delta\varphi = \varphi_{\text{pl}} - \varphi_\text{A} > 0$, the electron current to the anode is

$$I_{e\text{A}} = (1/4)\,e\bar{n}_e\,v_e\,S_\text{A}\exp\left(-e\,\Delta\varphi/T_e\right),$$

where $\bar{v}_e = \sqrt{8T_e/\pi m_e}$ is the mean thermal speed of the electrons. If we neglect the ion currents to the cathode and anode, then it is possible to use the approximation $I_{e\text{K}} \approx I_{e\text{A}}$, so that

$$\Delta\varphi \simeq (T_e/e)\ln(0.7\,S_\text{A}/S_\text{K}).$$

It is clear from this expression that in order to obtain $\Delta\varphi < 0$ it is necessary to make the anode area smaller than the cathode area.

Since the electrons are magnetized and can depart the plasma and go to the parts of the vessel walls that are at the anode potential primarily through the magnetic slits in the pole regions of the linear magnets, the effective anode area is roughly given by $S_\text{A} \approx L_\text{M}\delta_\text{M}$, where L_M is the overall length of the magnets and δ_M is the width of the magnetic slit for the electrons. A complete theoretical analysis of the dependence of δ_M on the field strength and plasma parameters has not yet been made; however, several experiments [56] show that δ_M may be close to $4\rho_h$, where $\rho_h = \sqrt{\rho_e\rho_i}$ is the hybrid Larmor radius for B_z at the wall. In our case, $B_z = 9 \cdot 10^{-2}$ T and for $T_e \approx 5$ eV and the probable $T_i \approx 0.4$ eV, δ_M is about 0.12 cm, i.e., $S_\text{A} \approx 70$ cm$^2 > S_\text{K}$.

S_A could be reduced by reducing L_M through increasing the step size d_M between the magnets; however, then the size of the transition region between the wall and the uniform plasma is increased, and this leads to a reduction in η and to the need to increase the vessel size. Another way is to use stronger magnets made, for example, of samarium–cobalt alloy, which could raise the magnetic field on the inner wall of the discharge vessel to 0.2-0.3 T.

Since the confinement time increases with the ratio of the plasma volume to the ion loss area, a simple way of increasing the H_1^+ fraction is to increase the vessel volume and reduce the possible ion loss area. Experiments show that increasing the magnetic field at the vessel wall from 0.12 to 0.27 T raised the H_1^+ fraction from 80 to 90% for an emission current density of 0.25 A/cm^2 [57]. In addition, predissociation of the molecular hydrogen fed into a duopigatron has been used at Oak Ridge to obtain a substantial increase in the H_1^+ fraction [49].

In order to enhance the proton content and improve the emission uniformity so-called magnetic filters have been used in an ion source with an edge magnetic field at Berkeley [57]. On the whole, however, both of these methods made construction of the discharge vessel much more complicated and in the second, reduced the gas efficiency. Magnetic filters have been used in the discharge vessels of negative ion sources.

3.4. Ion-Optical System and Neutralizer

Three-electrode ion-optical system. Although several different gaseous discharge plasma emitter systems have been developed, all ion sources employ a common type of ion-optical system, a multiaperture, three- or four-electrode electrostatic system for extracting, accelerating, and shaping the ion beams [2, 43].

The simplest three-electrode cell for a slotted ion-optical system is illustrated in Fig. 3.13. The profiles and dimensions of the apertures that form the ion-optical system are optimized in calculations of the parameters of an elementary beam (beamlet). Each elementary cell functions independently. The total ion current is the sum of the separate beamlets, whose directions and divergence angles determine the geometry of the composite beam. In optimizing the ion-optical system the problem is to obtain a divergence for the envelope of the composite beam of no more than ±1-2° for beamlet divergence angles in the range ±0.5°.

As an example we consider a typical three-electrode ion-optical system for an ion source with an edge magnetic field which was constructed as a separate component consisting of extracting (emitting), accelerating, and grounding electrodes mounted in an insulating assembly (unit II in Fig. 3.6). Each electrode consists of five grids with 12 slots in each. The size of the slots in the extraction electrode is 0.2×12 cm^2. The slots are located on a 12×25 cm^2 surface and form an overall emitting surface of 144 cm^2.

The geometry of a slotted cell in the ion-optical system is shown in Fig. 3.13 which also shows the dependence of the beamlet divergence angle perpendicular to the slot ($\theta_\perp$) on the reduced emission current density. The minimum divergence angle is $\theta_{\perp 0} \cong \pm 1.2°$ The divergence of the beam in the direction along the slot is about ±0.5°.

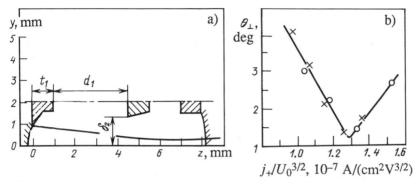

Fig. 3.13. The geometry of a slotted cell in an ion-optical system (a) and the dependence of the beam divergence angle in the direction perpendicular to the slot ($\theta_\perp$) on the reduced ion current density $j_+/U_0^{3/2}$ for $t_1 = 1$ mm, $d_1 = 3.3$ mm, and $\delta_2 = 1.2$ mm (b).

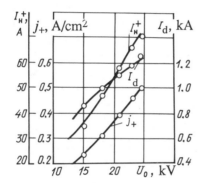

Fig. 3.14. The dependence of the optimum values of I_{H^+}, j_+, and I_d on the accelerating voltage U_0. (The geometry of the ion-optical system has been designed for production of 25- keV beams.)

The beam is focused in the direction along the slots by bending the grids to a radius of about 2 m, and in the direction perpendicular to the slots by shifting the accelerating electrode grids relative to the extraction electrode grids. The grids closest to the middle are shifted by about 150 μm, and the outermost grids, by 300 μm. The focal distance obtained from the relative displacement of the grids is given by [48]

$$F \simeq 1.5\, y_0\, d_{\text{eff}}/\Delta y,$$

where y_0 is the distance from the shifted grid to the central grid, Δy is the grid displacement, $d_{\text{eff}} = t_1 + d_1 + \delta_2$ is the effective length of the accelerating gap, t_1 is the thickness of the extraction electrode, d_1 is the distance between the extraction and accelerating electrodes (the length of the accelerating gap), and δ_2 is the half width of the slot in the accelerating electrode (linear dimensions in cm) (see Fig. 3.13).

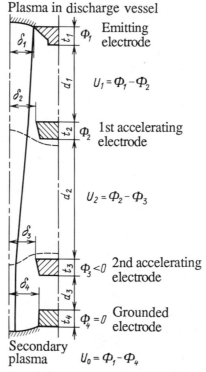

Fig. 3.15. The geometry of a four-electrode ion-optical system and the potential distribution on the electrodes.

Figure 3.14 shows some typical parameters, including the optimum beam current from the source I_{H^+}, the current density $j_+ = I_{H^+}/S_{em}$, and the discharge current I_d for which the divergence angle $\theta_\perp$ is minimal, as functions of the accelerating voltage U_0. The ion source with an edge field yields a hydrogen ion beam with a current of up to 70 A, the optimum values of j_+ for different U_0 agree well with the calculated values, and the relationship between j_+ and I_d is in agreement with data from probe measurements. The average beam divergence in the multislot ion-optical system exceeds the value obtained from a study of the focusing properties of a single-slot cell and is about $\pm 1.5°$. The increased divergence is caused mainly by imprecise fabrication and positioning of the electrodes in the ion-optical system.

The beam focusing described above ensures efficient propagation of a fast atom beam (after the charge exchange cell) through the inlet duct of a tokamak. The beam size (to a level of $1/e$) at a distance of about 1.4 m from the ion-optical system is 6×22 cm^2, smaller than the initial size. The mean power density in a 25-keV fast atom beam reaches 5 kW/cm^2 and on the beam axis it exceeds 10 kW/cm^2.

On the whole, the extraction and formation of beams with three-electrode ion-optical systems has been studied fairly well [2]. These ion-optical systems make it possible to efficiently form hydrogen ion beams with emission current densities j_+ of up to 0.5 A/cm^2 at energies eU_0 of up to 50 keV, where the geometric design of the ion-optical system and specified accelerating voltage U_0 correspond to an optimum current density $j_{+,0}$ at which the beam divergence is minimal. Experiments [2] show that $j_{+,0} \cong 4 \cdot 10^{-8} U_0^{3/2}/d_{eff}^2$.

In order that j_{+0} be as great as possible for a given accelerating voltage, it is necessary to reduce the length d_1, but there is always some limit $d_{1,min}$ below which the source stops operating normally because of electrical breakdown in the ion-optical system. Similarly, for every value of d_1 there is a limit to the accelerating voltage U_{max} and, therefore, a limit to the optimum ion emission current density. Experiments with different sources at the Kurchatov Institute, Culham, and Berkeley [2] show that for energies up to about 50 keV the relation between U_{max} and d_1 is fairly well described by $U_{max}(kV) = 80d_1^{0.8}$.

We now consider how the limiting current density $j_{+,max}$ depends on the particle energy. The thickness of the extraction electrode can hardly be made less than 1 mm for reasons of heat transfer and mechanical strength, and the half width δ_2 of the slot should be roughly $0.3d$. Then, assuming that the dependence of U_{max} on d is the same when extrapolated into the region above 50 kV, we find that: (1) the function $j_{+,max} = f(U)$ has a maximum in the region of $U \approx 20$ kV; (2) $j_{+,max}$ is roughly proportional to $1/U$ in the region $U > 100$ kV; and (3) in the region $U < 10$ kV d begins to be determined mainly by the thickness t_1 and as U is reduced the value of $j_{+,max}$ falls [2].

In order for an ion source to operate stably without breakdown over a long period at a given voltage, the length of the accelerating gap must be increased above the minimum; that is, the working value of $j_{+,0}$ must apparently be equal to $0.8j_{+,max}$. Thus, a three-electrode ion-optical system can be used to obtain 40-keV hydrogen ion beams with a current density of up to 2.2 A/cm^2, but at 100 keV $j_{+,0}$ falls to roughly 0.5 A/cm^2. It is therefore necessary to increase the plasma emitting surface area substantially in order to obtain the required ion current from the source.

Four-electrode ion-optical systems with two-stage ion acceleration [57] are intended for production of ion beams with high current densities and energies above 100 keV. In the first accelerating gap the potential difference should be about 30 kV, which makes 0.5 A/cm^2 feasible, and the remaining voltage is applied to the second accelerator gap.

Besides this possibility of obtaining beams with high current densities, multielectrode ion-optical systems with several accelerator stages have another positive characteristic of great significance in the development and utilization of sources with pulses longer than 1 s. Modeling of the distribution of fluxes of secondary charged particles in multielectrode ion-optical systems [58] shows that it is possible to redistribute these fluxes so that the loading of the electrodes will be in equilibrium and at minimal levels by selecting the geometry of the electrodes and the potential distribution on them.

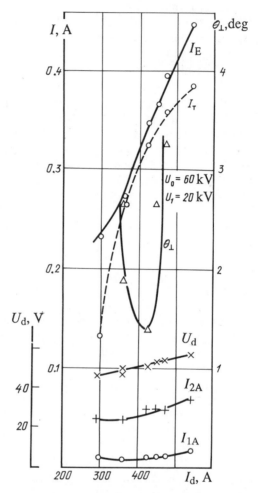

Fig. 3.16. The discharge current dependence of the discharge voltage U_d, the load
currents at the extraction (I_E) first accelerating (I_{1A}), and second accelerating (I_{2A}) electrodes, the calorimetrically measured beam
current I_T at a detector, and the beam divergence angle $\theta_\perp$ for $d_1 = 2.85$ mm and $d_2 = 6.5$
mm.

The focusing properties of a four-electrode ion-optical system depend on far
more parameters than do those of a three-electrode system. Let us consider a specific example [48] of beam formation in a four-electrode ion-optical system cell
(Fig. 3.15). The dimensions of the slots in the electrodes are $2\delta_1 = 2$ mm, $2\delta_2 =$
2.5 mm, $2\delta_3 = 2.5$ mm, and $2\delta_4 = 3$ mm, while the thickness of all the electrodes is
1 mm. During the experiment the discharge voltage and current were measured, as

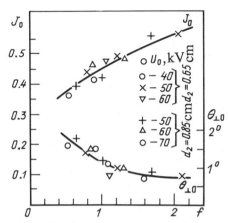

Fig. 3.17. The dependence of the minimum divergence $(\theta_{\perp 0})$ and the corresponding optimum reduced current density J_0 on the ratio of the electric field strengths in the second and first accelerating gaps, $f = E_2/E_1$.

well as the current loads at the rectifiers supplying the extraction electrode and the first and second accelerating electrodes, the beam current, and the beam profile at a detector target located 2 m from the source, from which the beam width and divergence were obtained. Typical dependences of these parameters on the discharge current I_d are shown in Fig. 3.16. For the optimum discharge currents at which the beam divergence is minimal, the beam current to the target (I_T) differs negligibly from the high-voltage rectifier current (I_E) to the extraction electrode. From the set of such dependences obtained for different values of the total voltage U_0 and the voltage on the first accelerating gap (U_1), it is possible to construct the dependence of the beam divergence $(\theta_{\perp})$ on the emission current density j_+ for different values of d_2, as well as on the relative current density $J = j_+/j_1$. Here j_1 is the current density calculated from the parameters of the first accelerating gap according to

$$j_1 = 4 \cdot 10^{-8} \, U_1^{3/2}/(d_1 + t_1)^2, \text{ A/cm}^2,$$

where U_1 is in volts, and d_1 and t_1 in cm (see Fig. 3.13a). For $d_1 = 0.285$ cm and $U_1 = 10, 15$, and 20 kV, $j_1 = 0.25, 0.46$, and 0.70 A/cm² respectively. Each combination of U_0, U_1, and d_2 has a corresponding optimum current density $j_{+,0}$ at which the beam divergence is mimimal $(\theta_{\perp 0})$. Here $j_{+,0}$ depends mainly on U_1 and the angle $\theta_{\perp 0}$ and the optimum $J_0 = j_{+,0}/j_1$ depend noticeably on the ratio $P = U_2/U_1$ of the voltages across the first and second accelerating gaps and on the ratio $g = d_2^{\text{eff}}/d_1^{\text{eff}}$ of the effective lengths of the accelerating gaps, where we may take $d_1^{\text{eff}} = t_1 + d_1 + t_2/2$ and $d_2^{\text{eff}} = t_2/2 + d_2 + t_3/2$. The generalized parameter on which both $\theta_{\perp 0}$ and J_0 depend is the ratio of the electric field strengths $f = E_2/E_1 = P/g$

(Fig. 3.17). As f is increased, J_0 rises and $\theta_{\perp 0}$ falls, which is explained by the shape of the equipotentials near the slot in the first accelerating electrode. In fact, near the second accelerating electrode the equipotentials are always convex toward the grounded electrode, so that a diverging lens is formed at this location. Near the first accelerating electrode when $f > 1$ the equipotentials are convex toward the extraction electrode, which leads to focusing of the beam (a converging lens), while when $f < 1$ they are convex toward the second accelerating electrode and a diverg­ing lens is formed. Therefore, when $f > 1$ the beam may even be parallel in the first accelerating gap, and the effect of the diverging lens near the second accelerating electrode is compensated by the converging lens near the first accelerating electrode, so that as it leaves the cell the beam is comparatively wide. When $f < 1$ the action of the two diverging lenses should be compensated by the strong convergence of the beam in the first gap, i.e., in this case the extracted current density should be lowered and the curvature of the plasma meniscus increased. Clearly the increase in $\theta_{\perp 0}$ is also related to the greater curvature of the plasma meniscus, since then the plasma shape may deviate substantially from cylindrical and cause noticeable aberration.

In a four-electrode ion-optical system, therefore, the best conditions for beam extraction and focusing are attained when the electric field strength in the second accelerating gap is greatest. In the system that was tested the minimum beam divergence was ±0.8° (at a level of $1/e$) for an emission current density of 0.4 A/cm².

Neutralizer. The scheme for charge exchange of positive hydrogen ions on hydrogen itself (for conversion of the ion beam into a beam of fast atoms) is recognized as the simplest from a technical standpoint and is used in most modern injectors. In order to ensure yield fractions of H^0 and H^+ that are close to equilibrium, the integrated thickness of the target must be on the order of $0.5\text{-}1.0 \cdot 10^{16}$ cm⁻². The neutralizer channel is directly adjacent to the ion source, so that all the gas flowing out of the ion source forms a target in the charge exchange channel. This considerably shortens the path length for an intense beam of charged particles and there is no need to attain high gas efficiency in the ion source (above 50%).

In order to make optimal use of the gas flowing out of the source, every ion source abuts its neutralizer channel. This enhances the flow resistance for molecular hydrogen and, thereby, the thickness of the charge exchange target. Designs have been examined in which the channel is further subdivided into separate sections and the channel walls are cooled by liquid nitrogen in order to lower the temperature of the neutral gas. The neutralizer channel walls also serve as a magnetic shield from the stray fields of a thermonuclear experiment, since the magnetic field strength inside the channel must not exceed a few times 10^{-4} T in order to avoid increasing the transverse velocity spread of the neutrals that are formed.

Chapter 4

MECHANISMS AND THEORY OF
ION SPUTTERING OF METALS

4.1 Ion Sputtering and Ion Implantation in Metals
and the Problem of Improving
the Properties of Metal Surfaces

The bombardment of solids by fluxes or beams of fast (with energies of 10 eV to 100 MeV) ions, atoms, or atomic nuclei induces a very complicated interaction. In order to study this interaction, one usually studies isolated processes or their manifestations. As it approaches a solid surface, a flux of ions pulls electrons out of the surface and is converted into a flux of fast atoms. In collisions with atoms on the surface layer, part of the fast atom flux is scattered, with changes in its direction of motion, and is reflected from the surface. Some of the incident atoms knock out or sputter atoms from the surface layer of the material and impurity atoms of residual gases, left over from evacuation of the volume, that have been adsorbed by the surface. Besides individual sputtered atoms and ions, the fast atomic particles also eject molecules and multimolecular compounds (clusters). Part of the potential and kinetic energy of the ions is lost through emission of secondary electrons and electromagnetic radiation in the form of light and x-ray photons.

The bulk of the incident atomic particles enter the solid material and collide with its atoms, creating primary ejected atoms, which in turn create secondary ejected atoms, etc. As a result, a cascade of atomic collisions develops. The number N_d of atoms in the target that are displaced from their positions in the crystal lattice of a material depends on the energy E_1 of the incident ions and the energy E_d required to displace an atom from a lattice point. A relatively small fraction of the displaced target atoms escape from the surface, leading to sputtering. If an atom is knocked out of a lattice point and the latter remains free, a vacancy is produced. An atom in an interstice and a vacancy are known as a Frenkel pair. As a consequence of multiple collisions and scattering events on target atoms, the incident atomic particles lose their electrons and may become multiply charged ions. Because of energy loss through elastic (or nuclear) and inelastic (or electronic) collisions, the bombarding particles lose their energy and are implanted in the solid body with a spread in their path lengths (total, or along the length of the trajectory, and projected, or along the projection of the total path in the direction of incidence of the ions on the surface).

Under certain conditions blisters, whose tops break and open up, are formed on irradiated metal surfaces. In addition flaking of the surface occurs.

The grain boundaries of all kinds of solids (metals, alloys, minerals, synthetic fibers) sputter more rapidly under ion bombardment than the grains themselves; hence, the structure of these materials is exposed by ion etching.

Under certain conditions films consisting of the residual gases from the vacuum system and of bombarding atoms are formed on the surface of a solid during ion bombardment.

Ion bombardment stimulates sorption and desorption, as well as diffusion of implanted particles into the depth of a target.

During ion bombardment cones, depressions, ridges and trenches, crystal-faced pits, and multilayer dislocation rings may be formed on the surface of poly-crystals and monocrystals.

Some of the processes that accompany the interaction of atomic particles with solids have already found wide application in industrial technology. For example, ion (cathode) sputtering and ion implantation are successfully used in the manufacture of semiconductor devices and integrated circuits for microelectronics.

The mechanisms of ion sputtering, radiation damage, energy loss, the distribution of implanted atoms along their paths, reflection and scattering of ions, radiation blistering, and other processes are the subject of many thousands of papers. These studies have mainly dealt with semiconductors, metals, and dielectrics.

A comparatively small number of studies, however, have been performed on steels and cast iron with the purpose of making them corrosion resistant by means of bombardment with atomic particles. Of the total of all mined metals, 94% is iron. Iron is the material basis, the "skeleton," of our civilization. The demand for steels is constantly and rapidly increasing. Thus, over the 33 years from 1890 to 1923, 1.8 billion tons of steel and cast iron were produced in the whole world. Over the last ten years the same amount of the metal has been produced in the USSR. The planned output in 1985 of 109.4 million tons of rolled ferrous stock and 19.7 million tons of steel pipe does not fully meet the requirements of the Soviet economy. The problem is to reduce the use of the metal per unit GNP to 2.5% and the rate of consumption of ferrous rolled stock in machine building and metal processing to 3.7%. One of the reasons for the shortage of metal is the vast loss owing to corrosion. According to the calculations of one American institute, 40% of the above amount of metal production over 33 years (718 million tons) was irreversibly lost because of corrosion. Data from the Third International Congress on Corrosion in Metals indicate that the annual loss of metal by corrosion is 10-12% of of production. Direct loss by corrosion is roughly $13 billion per year in the entire world. This means that at any given moment, one in eight or ten enterprises involved in the mining and enrichment of ore (mines and coke, metallurgical, rolling, and machine building factories) works to recover losses from corrosion. This is distinctly clear from data of the Central Statistical Administration of the USSR [59].

Nevertheless, although direct losses from corrosion are extremely large, indirect losses from it are much greater. The operating lifetime of many tens and hundreds of millions of machines and structures is reduced. Machines, structures, and buildings, whose value is much greater than that of the metal in them, are destroyed. Chemical, power, and pipeline equipment stands idle. (In the Soviet Union tens of thousands of kilometers of major oil pipelines are being overhauled.) The productivity of oil and gas pipelines drops as a result of contamination by corrosion products. Heat exchange is reduced in the boiler furnaces of thermal electric power stations, in nuclear power stations, and in internal combustion engines. The cost of metal in machines, pipelines, and buildings is certainly raised. Accidents occur in machines and buildings [60]. Many methods for battling with corrosion have been developed, beginning with the alloying of steels and ending with various methods for applying coatings to steel and cast iron. The fundamental shortcomings of these methods for enhancing corrosion and wear resistance are that alloying the entire volume of metal leads to a substantial (up to a factor of 10) rise in the cost of stainless steel compared to low-alloy steels and that coatings adhere relatively weakly to metals. Since the corrosion resistance of an overwhelming majority of equipment is determined by the surface layer, ion doping may be used to create barrier layers out of inert gases or to obtain thin layers (50-80 nm) of stainless (chromium nickel) steels on iron or steel surfaces. Ion doping enhances the hardness, durability, plasticity, and fatigue lifetime of steels. Ion doping makes it possible to obtain new alloys or metastable solid solutions which cannot be obtained by the usual methods.

Cathode sputtering of two elements has made it possible to obtain superconducting Nb_3Ge with a record high superconducting transition temperature. This process makes it possible to obtain pure surfaces and thin films, and to mill, bore, grind, and polish surfaces with high precision and at rather high speeds.

Ion beam processing of solid surfaces opens up great possibilities for technology. It might be said that ions are "knocking" at the door of technologists. In order to describe and understand these possibilities it is first necessary to become acquainted with the most important aspects of the interaction of atomic particles with metals. The limited size of this book makes it impossible to examine every such process in detail. Hence, we shall discuss the basic concepts, mechanisms, and theory of the sputtering of metals by ions, the most important interaction process.

4.2. Rate of Ion (Cathode) Sputtering

Ion sputtering or, more precisely, the sputtering of solid and liquid substances by fast atomic particles (ions, atoms, molecules, clusters, atomic nuclei, electrons, and neutrons), involves the sputtering of these materials in the form of atoms, positive and negative ions and clusters (multiatomic or molecular complexes). The bombardment of solids and liquids by atomic particles leads to sputtering of atoms and molecules of the gases dissolved in them as well as to the ejection of atoms implanted during bombardment by the atomic particles. This process is known as physical sputtering.

During irradiation of solids by chemically active ions, radicals, and atoms (for example, by hydrogen or halogen molecules), the mass of samples decreases both because of physical sputtering mechanisms and because of the formation of volatile chemical compounds (hydrides, halides). The latter process is known as chemical, plasma chemical, or ion chemical sputtering.

The sputtering of solids by ions is characterized by the sputtering yield (sputtering ratio or sputtering coefficient) Y,[*] which is equal to the number of ejected atoms per ion (atoms/ion), and the sputtering rate v_s, which is defined as the thickness of target material removed per unit time (cm/s):

$$v_s = \frac{dx}{dt} = \frac{dm}{dt\,\rho\,cm^2} = \frac{M_2\,Y}{N_A\,q_i} = 1.04{\cdot}10^{-5}\,j_i\,YM_2/\rho, \qquad (4.1)$$

where M_2 is the mass number of a target atom in relative units, dm is the mass of sputtered material (g), and

$$Y = dm\,N_A\,e/(M_2 I_i\,t); \qquad (4.1a)$$

where $N_A = 6.02{\cdot}10^{23}$ mole^{-1} is the Avogadro number, $q_i = Z_i e$ is the charge of a sputtered ion, $e = 1.6{\cdot}10^{-19}$ C is the electronic charge, j_i is the ion current density (A/cm^2), ρ is the target density (g/cm^3), I_i is the ion current (A), and t is the duration of the ion bombardment (s).

Sometimes the sputtering rate is expressed as the mass of material sputtered from 1 cm^2 of surface per second, v_m (10^{-9} kg/cm^2{\cdot}s):

$$v_m = \frac{dm}{A_s\,dt} = \frac{M_2\,Y}{N_A\,q_i}\,j_i = 0.104\,j_i\,YM_2. \qquad (4.1b)$$

The ion current density produced by ion sources varies over wide limits, from 10^{-6} to 0.1 A/cm^2. The sputtering yield may vary over the range 10^{-4} to 10^1. For iron the ratio $M_2/\rho = 55.8/7.9 = 7.1$. Substituting these values in Eq. (4.1), we find that ion sputtering can be used to process any solid material at a rate of 10^{-14} to 10^{-4} cm/s.

---

[*]The most often used notation for the sputtering yield is S; but sometimes γ, K_s, and other notations are used.

TABLE 4.1. Threshold Energy E_t (eV) for Sputtering of Some Metals by Ions [66, 69]

Metal	Ion				Metal	Ion			
	H^+	D^+	$^4He^+$	Hg^+		H^+	D^+	$^4He^+$	Hg^+
Be	27.5	24	33	–	Fe	64	40	35	12
Al	53	34	20.5	–	Zr	–	–	60	12
Au	184	94	44	4	Mo	164	86	39	16
Ag	–	–	–	5	Cr	–	–	–	23
Cu	–	–	–	7	Ti	43.5	–	22	24
Pt	–	–	–	8	V	76	–	27	–
Ni	47	32.5	20	9	W	400	175	100	25
Co	–	–	–	11	Nb	–	–	–	27
					Ta	460	235	100	35

4.3. Dependence of the Sputtering Yield on the Ion Energy

The sputtering yield depends on the energy, charge, and mass of the ions, their angle of incidence upon the surface, and the ion dose, as well as on the nuclear charge, mass, and binding energy of the target atoms, the crystal structure, the cleavage plane of a monocrystal and its rotation direction, contamination of the surface by oxides and films, the roughness and porosity of the surface, the nature of the residual gas and its pressure, the sample temperature, and several other bombardment conditions [61-68]. We now examine the most important of these features of ion bombardment.

Sputtering sets in when the ion energy exceeds the threshold energy for sputtering. With the aid of extremely sensitive techniques for measuring the sputtering yield, Morgulis and Tishchenko, Stuart, Wehner, and others [62] have found experimentally that the sputtering yield is 10^{-4}-10^{-5} for ion energies of 3-30 eV. The threshold sputtering energy E_t depends on the ion and atom masses and on the binding energy of the atoms in the solid surface, which is determined by the heat of sublimation U_0 of the material. If we assume that an atom is sputtered by the transfer of momentum and energy to it from an incident ion, then the threshold sputtering energy can be defined as

$$E_t = U_0/\gamma_m, \qquad (4.2)$$

where

$$\gamma_m = 4M_1M_2/(M_1 + M_2)^2. \qquad (4.2a)$$

The experimentally determined threshold energies for sputtering of iron by Ne^+, Ar^+, Kr^+, Xe^+, and Hg^+ are 12-28 eV. Table 4.1 lists the threshold sputtering energy for a number of metals.

TABLE 4.2. Sputtering Yields of 316 and 304 Stainless Steels Bombarded by H^+, D^+, and He^+ Ions [66]

Energy, eV	Sputtering yields for the ions		
	H^+	D^+	He^+
100	–	$3.35 \cdot 10^{-3}$	–
150	–	$7.85 \cdot 10^{-3}$	$(5.9 \pm 1.0) \cdot 10^{-2}$
200	$2.73 \cdot 10^{-3}$	–	–
250	–	–	$7.09 \cdot 10^{-2}$
300	$3.64 \cdot 10^{-3}$	$1.72 \cdot 10^{-2}$	–
400	$5.5 \cdot 10^{-3}$	–	–
500	$8.67 \cdot 10^{-3}$	$(2.25 \pm 0.25) \cdot 10^{-2}$	$(8.6^{+0.7}_{-1.3}) \cdot 10^{-2}$
1000	$\underline{9.77 \cdot 10^{-3}}$	$\underline{2.6 \cdot 10^{-2}}$	$(10.5 \pm 2) \cdot 10^{-2}$
2000	$(8.8^{+0.5}_{-0.8}) \cdot 10^{-3}$	$(2.6^{+0.3}_{-0.4}) \cdot 10^{-2}$	$(1.5^{+0.1}_{-0.3}) \cdot 10^{-1}$
3000	$(9.1 \pm 0.3) \cdot 10^{-3}$	$(2.5 \pm 0.4) \cdot 10^{-2}$	–
4000	$7.83 \cdot 10^{-3}$	$2.18 \cdot 10^{-2}$	$\underline{(1.36 \pm 0.03) \cdot 10^{-1}}$
6000	$(5.0 \pm 0.35) \cdot 10^{-3}$	$(1.44 \pm 0.1) \cdot 10^{-2}$	–
8000	$1.6 \cdot 10^{-3}$	–	$(8.6^{+0.4}_{-0.6}) \cdot 10^{-2}$

Note: The maximum values of Y are underlined.

When the energy of Ar^+ and Hg^+ ions is raised from 20-30 eV to 100 eV, the sputtering yield of metals increases rapidly from 10^{-4} to 10^{-1}. The variation in the sputtering yield for a variety of metals obeys the relation [69]

$$Y = k(E - E_t)^3, \tag{4.3}$$

where k is a coefficient that depends on the properties of the target material. As the ion energy is increased further, the sputtering yield rises more slowly, reaches a maximum, and then begins to decrease.

An empirical formula has been proposed [70, 71] for finding the sputtering yield at reduced energies $E' \leq 20$ (relative to the threshold energy E_t') including the reflection of ions from the surfaces of metals and carbides:

$$Y = 6.4 \cdot 10^{-3} M_2 \, \gamma_m^{5/3} E'^{1/4} (1 - 1/E')^{7/2}, \quad E' < 20, \tag{4.4}$$

where
$$E' = E/E_t';$$

$$\left.\begin{array}{l} E_t' = U_0/\gamma_m(1 - \gamma_m), M_1/M_2 \leq 0.3; \\ E_t' = 8 U_0 (M_1/M_2)^{2/5}, \quad M_1/M_2 > 0.3. \end{array}\right\} \tag{4.5}$$

This formula describes well the measured sputtering yields for 11 metals, graphite, silicon, and the carbides B_4C, SiC, TaC, and WC irradiated by H^+, D^+, He^+, Ne^+, Ar^+, Kr^+, and Xe^+ ions (a total of 250 ion–target combinations). It can be used to estimate sputtering yields up to several keV.

The measured sputtering yields for the stainless steels that are candidates for the first wall material in thermonuclear reactors are shown in Table 4.2. These data show that for light ions the maximum of the sputtering yield as a function of the energy occurs at energies of 1-4 keV. During bombardment of metals by ions with medium and high relative atomic masses, this maximum becomes broader and shifts toward energies of tens and hundreds of keV. The energy dependence of the sputtering yields for energies of 1-100 keV is well known for copper, silver, molybdenum, and several other metals [62-64].

4.4. The Theory of Sputtering by Ion Bombardment in the Linear Range of Atomic Collision Cascades

During the more than 130 years that the physical phenomenon of ion sputtering has been studied, quite a few hypotheses have been advanced on the nature of and mechanism for this process: a chemical interaction similar to electrolysis in gases, "electrical" evaporation, instantaneous evaporation from a microscopic particle of the material heated by an ion, transfer of momentum from an ion to a target atom, a chemical interaction with the formation of hydrides, and an explosion of gas in the material. During the period 1926–1946 the most widely accepted theory was the thermal theory of cathode sputtering proposed by von Hippel and developed by others. Experiments with ion beams at energies of 10–50 keV, however, showed that the angular distribution of the sputtered atoms differed considerably from the isotropic (cosine) distribution observed for evaporating atoms and for sputtered atoms at ion energies of about 10 keV [62]. Further research on the sputtering of monocrystalline targets at energies of from 50 eV to 1200 keV has shown that atoms are emitted primarily in the directions with close packing of the atoms in atomic rows. The measured average energies and the energy spectra of sputtered atoms are considerably higher than the average energies and spectra of evaporating atoms.

A number of theories of the sputtering of polycrystalline, amorphous, and monocrystalline solids over a wide energy range have been developed in the last 30 years based on an impulse mechanism for cathode sputtering and the theory of radiation damage in solids. Of these theories the theory of sputtering of amorphous and polycrystalline targets developed by Sigmund [72] has gained wide acceptance. It is assumed that sputtering occurs as the result of the direct knocking out of an atom by an ion and the development of linear and nonlinear sequences or cascades of atomic collisions produced by ions with a given energy, mass, and current density. The sputtering yield is calculated assuming random scattering and slowing down of the ion in an infinite medium. An integrodifferential equation based on the Boltzmann transport equation is obtained and solved to obtain formulas for the sputtering yields in the reverse and forward directions along the beam path. An equation for the effective escape depth of the sputtered atoms is obtained. For copper this depth is 0.48 nm.

In the elastic atomic collision region (for example, for 1-MeV protons incident on a gold target)

$$Y = \Lambda \alpha \, N S_n (E),$$ (4.6)

TABLE 4.3. Reduced Nuclear Slowing-Down Cross Section $s_n(\varepsilon)$ for a Thomas–Fermi Interaction (from Linhard et al. [72])

ε	$s_n(\varepsilon)$	ε	$s_n(\varepsilon)$	ε	$s_n(\varepsilon)$	ε	$s_n(\varepsilon)$
0.002	0.120	0.04	0.311	0.4	0.405	10	0.128
0.004	0.154	0.1	0.372	1.0	0.356	20	0.0813
0.01	0.211	0.2	0.403	2.0	0.291	40	0.0493
0.02	0.261			4.0	0.214		

TABLE 4.4. Experimentally Measured Sputtering Yields Y_{exp} for a Number of Metals and the Values Y_{theor} Calculated using Sigmund's Theory (in parentheses) for 45-keV Inert Gas Ions [73]

Metal	U_0, eV	Z_2	Sputtering ratio			
			Ne^+	Ar^+	Kr^+	Xe^+
Pb	2.01	82	3.6	10.5 (29.7)	24.0 (44.4)	44.5 (74.6)
Ag	2.94	47	4.5	10.8 (12.7)	23.5 (20.1)	36.2 (24)
Sn	3.11	50	1.8	4.3 (12.5)	8.5 (19.5)	11.8 (24.9)
Cu	3.46	29	3.2 (3.7)	6.8 (6.7)	11.8 (11.9)	19.0 (15.5)
Au	3.79	79	3.6	10.2 (16.1)	24.5 (23.9)	39.0 (27.8)
Pd	3.87	46	2.5	5.3 (9.24)	10.5 (14.1)	14.4 (18.1)
Fe	4.29	26	1.3	2.3 (5.15)	4.0 (8.92)	4.9 (11.7)
Ni	4.43	28	1.4	3.5 (5.35)	5.6 (9.01)	7.6 (11.8)
V	5.3	23	0.3*	1.0 (3.85)	1.7 (6.21)	1.9 (8.9)
Pt	5.82	78	1.9	5.3 (9.35)	11.3 (15.3)	16.0 (19.2)
Mo	6.82	42	0.6	1.5 (5.14)	2.7 (7.70)	3.8 (10.1)
Ta	8.06	73	0.7	1.6 (6.62)	3.1 (10.0)	4.0 (12.3)
W	8.70	74	1.0	2.3 (5.07)	4.7 (9.31)	6.4 (11.9)

where

$$\Lambda = 3/(4\pi^2 \, NC_0 \, U_0) = 4.2 \cdot 10^{14}/(NU_0). \tag{4.7}$$

Here Λ is a coefficient that describes the properties of the target material and the state of its surface; α is a factor that depends on M_2/M_1 and the angle of incidence and energy of the ion [64, 72];

$$S_n(E) = [1/(1-m)] \, C\gamma^{1-m} E^{1-2m} \tag{4.8}$$

is the slowing-down cross section for nuclear or elastic collisions; and $m = 0\text{-}1$.

TABLE 4.5. Comparison of the Ratios of the Theoretical Sputtering Yields Y_{theor} Obtained from Sigmund's Formula with the Experimental Values Y_{exp} Given in Table 4.4 (Y_{theor}/Y_{exp})

Metal	Ar$^+$	Kr$^+$	Xe$^+$	Metal	Ar$^+$	Kr$^+$	Xe$^+$
Pb	2.8	1.85	1.7	Ni	1.5	1.6	1.6
Ag	1.2	0.85	0.66	V	3.85	3.65	4.7
Sn	2.9	2.3	2.1	Pt	1.8	1.35	1.2
Cu	0.99	1.0	0.815	Mo	3.4	2.85	2.65
Au	1.6	0.975	0.71	Ta	4.15	3.2	3.1
Pd	1.75	1.35	1.25	W	2.2	2.0	1.85
Fe	2.2	2.2	2.4				

Note: Significant deviations of the theoretical from the experimental values are underlined.

Sigmund obtained several formulas for different types of atomic collisions and various energy ranges.

a) For energies below 1 keV when the ions are incident along the normal to the surface,

$$Y = (3/4\pi^2)\, a\, T_m/U_0 = 0.076\, a\, T_m/U_0, \tag{4.9}$$

where $T_m = E4M_1M_2/(M_1 + M_2)^2$.

For Be, Si, Cr, Ni, Cu, Ge, Ru, Rh, Pd, Ag, Ir, Pt, and Au this formula is in good agreement with the experimental data of Wehner et al. For Ti, V, Zr, Nb, Hf, Ta, Th, and U the measured sputtering yields are a factor of 2 smaller than the calculated values. It should be noted that in Wehner's experiments the flux of ions bombarding the target was measured electrically. A milliammeter measured the combined current of ions and secondary electrons. Thus, the experimental values of the sputtering yield were represented in the form $Y/(1 + \gamma)$, where γ is the secondary ion–electron emission coefficient. This coefficient depends strongly on the purity of the target surface. This must be taken into account in using the numerous systematic studies of cathode sputtering undertaken by Wehner and his co-workers. In the meanwhile, some authors do not take the correction for the secondary emission coefficient, which may be large, into account.

b) For the keV energy range and medium ion masses Sigmund obtained

$$Y = 4.2 \cdot 10^{14}\, a S_n(E)/U_0, \tag{4.9a}$$

where

$$S_n(E) = 4\pi Z_1 Z_2\, e^2\, a_{12}\, [M_1/(M_1 + M_2)]\, s_n(\epsilon). \tag{4.10}$$

Here

$$\epsilon = \frac{M_2 E / (M_1 + M_2)}{Z_1 Z_2 e^2 / a_{12}} \; ; \qquad (4.11)$$

$$a_{12} = 0.885 \, a_0 (Z_1^{2/3} + Z_2^{2/3})^{-1/2}; \qquad (4.12)$$

and $s_n(\epsilon)$ is a universal function tabulated in Table 4.3.

A comparison with experimental data for copper, silver, palladium, gold, cadmium, zinc, germanium, and silicon bombarded by inert gas ions generally yields good agreement. However, in a number of cases ($Xe^+ \to Cu$, 50-150 keV; $Xe^+ \to Ag$, 20-400 keV; $Kr^+ \to Ag$, 30-80 keV) Eq. (4.9a) gives values that are 25-50% low. In other cases (Xe^+, Kr^+, Ar^+, $Ne^+ \to Pd$, $Ne^+ \to Ag$, $Ar^+ \to Au$) at ion energies of 5-500 keV it gives excessive values of the sputtering yield (see Table 4.4).

The theoretical values of the sputtering yields for vanadium, tantalum, molybdenum, tin, tungsten, iron, and lead exceed the experimental values by factors of 2-4, while for copper, gold, and silver there is good agreement between theory and experiment (see Table 4.5).

The theory of sputtering and slowing-down of charged particles in solids [74] is based on a Linhard power-law interatomic interaction potential

$$V(r) = Z_1 Z_2 e^2 a^{n-1} / (n r^n), \qquad (4.13)$$

where n is an exponent which varies from 5 for solid sphere collisions to 1 for Rutherford collisions. In this theory it is assumed that the first collision of an ion with a metal atom is weakly screened ($n = 2$) or of the Rutherford type, while subsequent collisions of displaced atoms are similar to collisions of hard spheres ($n = 5$). Proceeding from these assumptions, expressions were obtained for the maximum sputtering yield Y_{max}, for the maximum corresponding ion energy E_{max}, and for a universal yield-energy curve:

$$Y_{max} = K_0 \, N \pi \, a^2 / (\epsilon_d / \gamma_m). \qquad (4.14)$$

Here

$$K_0 = K(\nu_p + \nu_r), \qquad (4.15)$$

where K is a constant that characterizes the material; ν_p and ν_r are quantities that depend on the type of energy transfer from the incident particle to the target atoms; a is the effective atomic radius; ϵ_d is the reduced displacement energy, which is related to the reduced threshold energy by

$$\epsilon_d = \gamma_m \, \epsilon_t; \qquad (4.16)$$

$$Y / Y_{max} = (1/0.36) \, (n \, \lambda_n / 2) \, (\epsilon_d \, \epsilon / \gamma_m \, \epsilon_m)^{1/2 \, - \, 1/n}, \qquad (4.17)$$

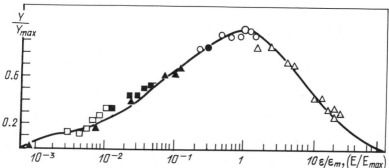

Fig. 4.1. Dependence of the normalized sputtering yield for copper on the normalized energy of Ar^+ ions calculated using Eq. (4.17) for $Y_{max} = 7$, $E_{max} = 42$ keV, and $\varepsilon_d/\lambda = 3.5 \cdot 10^{-5}$ on the basis of experimental data by selecting values of $n = 1$-5 and $\lambda_n = 0.0064$-23 [74]: □) Wehner et al.; ■) Weijsenfeld; ○) Almen et al.; ●) Kanaya et al.; ▲) Keywell; △) Dupp et al.

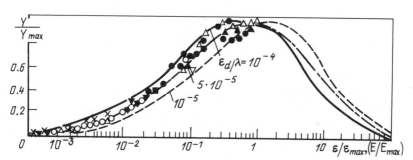

Fig. 4.2. Universal dependence of the sputtering yield for a number of metals on the normalized energy of Ar^+ ions calculated using Eq. (4.17) with $\varepsilon_d/\lambda = 10^{-5}$ (1), $5 \cdot 10^{-5}$ (2), and 10^{-4} (3) [74] on the basis of the following experimental data:

Notation	Metal	Y_{max}	E_{max}, keV	$\varepsilon_d/\lambda, 10^{-5}$	Authors
●	Ag	12	63	1.7	Keywell, Almen et al.
▲	Pd	6	62	3.1	M. I. Guseva
△	Sn	5	66	1.96	Almen, Bruce
○	Ni	6	41.8	4.23	Laegreid
▼	Fe	5	39	4.25	"
×	Hf	4	96	3.03	Wehner
▽	Ta	4	97.7	3.56	"

where λ_n is a scale factor given by

$$n \lambda_n/2 = (1/n) \left[(3n - 1)/8n^2 \right]^{1/n};$$

(4.18)

ε is the reduced energy [Eq. (4.11)]; $\varepsilon_{max} = 0.3$-0.4 is the maximum reduced energy for different atomic models; and γ_m is given by Eq. (4.2a).

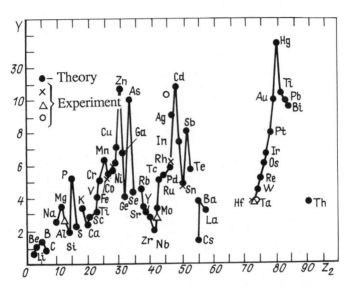

Fig. 4.3. Dependence of the peak sputtering yield with Ar⁺ given by Eq. (4.19) (•) for a number of elements on the atomic number of the sputtered substance compared with experimental data (O, ×, Δ [74]).

In order to obtain a universal dependence of the sputtering yield on the ion energy, the energy dependences of the sputtering yields normalized to the maximum sputtering yield and of the energies at which the sputtering yield reaches its peak were calculated for a variety of metals bombarded by Ar⁺ ions (Figs. 4.1 and 4.2).

4.5. Peak Sputtering Yields in the Linear Range of Atomic Collision Cascades

This question is of practical importance and research on it involves several aspects or trends. It is desirable to know the ion energy at which the sputtering yield is greatest. It is interesting to determine the material which has the maximum sputtering yield for a given ion and to find the type of ion which has the maximum possible sputtering effect. It is necessary to examine other factors on which the sputtering yield may be expected to depend strongly.

Japanese researchers [74] have found an answer to several of these questions. Assuming that the threshold sputtering energy is proportional to the sublimation energy (4.2), it follows that

$$Y_{max} = K_0 N \pi a^2 Z_1 Z_2 e^2 (M_1 + M_2)/(M_2 U_0) = 6.5 \cdot 10^{-10} N \pi a^2 / \epsilon_t, \qquad (4.19)$$

where $6.5 \cdot 10^{-10}$ cm is an empirical constant. This provides a theoretical confirmation of the property (Fig. 4.3), discovered in 1961 by Laegreid and Wehner [62],

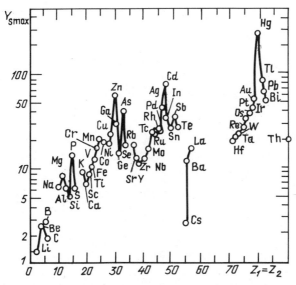

Fig. 4.4. Dependence of the maximum values of the "self-sputtering" yield on the atomic number of the ion and target given by Eq. (4.21) [74].

TABLE 4.6. Comparison of Experimental ($E_{max,exp}$) and Theoretical ($E_{max,theor}$) Values of the Ion Energies at Which the Maximum Sputtering Yield Is Observed

Energy, keV	Target				
	Ag	Cu	Au	Pd	Sn
$E_{max,exp}$	60	40	100	>50	$\gg 60$
$E_{max,theor}$	63.6	42.2	105.8	62	66.8

that the atom yield is greatest for elements with filled $3d$, $4d$, and $5d$ shells. The variation in the heat of sublimation with atomic number has a similar periodicity.

The energy at which the sputtering yield reaches a maximum is given by

$$E_{max} = \epsilon_m Z_1 Z_2 e^2 (M_1 + M_2)/(aM_2) . \qquad (4.20)$$

Good agreement is observed between theory and experiment for argon ions (Table 4.6).

It might be supposed that the "self-sputtering" of materials by ions of the same material would yield the highest sputtering yields. Figure 4.4 shows the results of theoretical calculations according to the formulas[*]

---

[*]Kanaya et al. [74] give an incorrect value of the coefficient (10^{-21}) in their Eq. (19) and Fig. 9.

$$Y_{s\,\text{max}} = 1.14 \cdot 10^{-23}\, NZ^{4/3}/U_0 \tag{4.21}$$

and

$$E_{s\,\text{max}} = 34.7\, Z^{7/3}. \tag{4.22}$$

This type of dependence was first found experimentally in 1961 by Almén and Bruce [62] for 27 metals at ion energies of 45 keV.

It is important to note that the theory predicts data that are needed in practical work (Table 4.7). The table shows, in particular, that the self-sputtering yields for the first wall of a thermonuclear reactor made of light elements (lithium, carbon, beryllium, boron) will be on the order of 1.7-2.7 atoms per ion for 160-190 keV ions, while for iron, nickel, niobium, and molybdenum walls the self-sputtering yield is expected to be on the order of 7-20 atoms per 280-340 keV ion.

The authors of the theory regard these qualitative relations as a first approximation. A more precise comparison of the theory with experimental data requires more accurate values of the parameters λ_n, ε_t, and α, which were chosen empirically for these calculations, since they do not depend on the theoretical assumptions.

Ryzhov and Shkarban [75] consider that the thickness of the "active layer," from which "quasistationary diffusion" of displaced atoms toward the surface takes place, may reach roughly 10 nm. This statement is based on an analysis and generalization of experimental data. It is also known from experiments with radioactive isotopes of gold by Patterson and Tolin [62] that atoms leave the layer from a depth of up to 8 nm.

A formula has been found [75] for the sputtering yield during bombardment by ions along the normal to the surface

$$Y = 0.26\,(\overline{\Delta E}/U_0^{1.4})\,[1 - B\exp(-0.1\,\lambda\, N^{1/3})], \tag{4.23}$$

where

$$\overline{\Delta E} = \pi^2\, A_0\, N\overline{m}/(\overline{m}+1)^2; \tag{4.24}$$

$\overline{\Delta E}$ is the average energy delivered to the layer of material calculated using a Firsov potential for the interatomic interaction potential; $\overline{m} = M_1/M_2$; A_0 is the constant in the Firsov potential given by

$$A_0 = \frac{3.05 \cdot 10^{-16}\, Z_1 Z_2}{(Z_1^{1/2} + Z_2^{1/2})^{2/3}}\, N^{2/3}; \tag{4.25}$$

$B = 1.8\overline{m}^{-0.135}$; and λ is the mean free path of an ion in the material, with

$$\lambda = (\pi R_{\text{int}}^2\, N)^{-1}, \tag{4.26}$$

where R_{int} is the interaction radius in the hard sphere collision model with a screening potential.

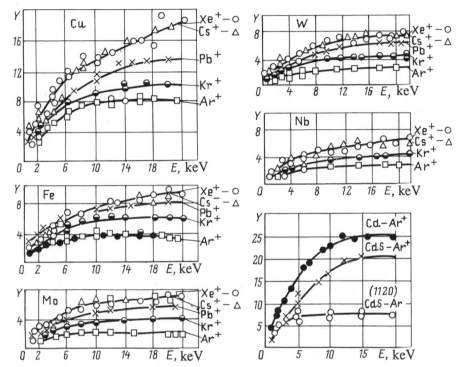

Fig. 4.5. Dependence of the sputtering yields of copper, iron, molybdenum, tungsten, niobium, cadmium, and cadmium sulfide on the energy of Ar^+, Kr^+, Xe^+, Cs^+, and Pb^+ ions [75].

Equation (4.23) yields an expression for the maximum sputtering yield,

$$Y_{max} = 0.26\ \overline{\Delta E}/U_0^{1.4}, \tag{4.23a}$$

which, however, is not necessarily equal to the actually attainable maximum value.

Figure 4.5 shows sputtering curves for several metals obtained by various authors. A comparison of the computed values of Y/Y_{max} with the experimental data for 60 ion–target atom combinations revealed a discrepancy within the limits ±16%. It should be noted that most measurements of the sputtering yield have an error of 10-30%.

The above formulas are complicated and often contain fitted parameters. Smith [76] has used the theory of Sigmund and the experimental data of various authors to choose two correction factors and obtain a formula which is suitable for engineering calculations:

$$Y = \frac{20}{U_0}\ Z_1^2 Z_2^2\ \frac{M_1}{M_2}\ \frac{E}{(E + 50\, Z_1\, Z_2)^2}. \tag{4.27}$$

TABLE 4.7. Calculated Maxima of the Sputtering and Self-Sputtering Yields for 58 Elements Irradiated by Argon and Xenon, Together with the Parameters Used in Eqs. (4.19)–(4.22) [74]

Target	Z_2	N, 10^{22} cm^{-3}	U_0, eV	$Y_{s\ max}$	Ar$^+$			Xe$^+$		
					a, 10^{-4} nm	Y_{max}	E_{max}, keV	a, 10^{-4} nm	Y_{max}	E_{max}, keV
Li	3	4.59	1.69	1.35	154	0.65	13.4	116	1.08	160
Be	4	12.3	3.48	2.58	149	1.26	14.7	114	2.32	169
B	5	13.7	5	2.7	147	1.38	16.3	113	2.55	181
C	6	11.3	7.41	1.87	144	1	18.3	112	0.88	191
Na	11	2.49	1.11	6.38	134	2.6	22.9	107	7.2	215
Mg	12	4.31	1.59	8.9	132	3.5	24.4	106	9.69	225
Al	13	6.04	3.26	6.5	130	2.54	25.2	105	7.74	226
Si	14	5.04	3.91	5	129	1.97	26.7	104	5.83	236
P	15	4.55	1.37	14	128	5.2	28.4	103	16.4	237
S	16	3.75	2.89	6	126	2.28	28.9	103	6.88	245
K	19	1.31	0.85	9	123	3.3	31.8	101	10.7	254
Ca	20	2.31	2.1	6.85	122	2.4	33.4	101	7.7	255
Sc	21	4.03	3.08	8.7	121	2.8	33.4	100	9.9	255
Ti	22	5.66	4.34	9.25	119	2.9	34.3	99	10.2	257
V	23	6.63	3.7	13.4	118	4.06	35.2	99	15.1	258
Cr	24	7.64	3.68	16.5	118	5.1	36.6	99	18.1	267
Mn	25	8.01	3.15	21.4	117	6.4	37.6	98	25.7	268
Fe	26	8.48	4.15	18	116	5.3	39	98	19.4	278
Co	27	8.9	4.38	18.1	115	5.65	39.9	97	19.9	279
Ni	28	9.03	4.41	19.9	115	6	41.8	97	20.6	280
Cu	29	8.52	3.56	24.6	114	7	42.2	96	25.1	285
Zn	30	6.48	1.19	58.2	113	16	43.5	96	59.8	290
Ga	31	5.12	2.27	25.1	112	6.7	44.2	95	25.3	291
Ge	32	4.53	3.77	14	112	3.7	45.2	95	13.7	294
As	33	4.61	1.32	42.5	111	11.1	46.4	95	40.7	299
Se	34	3.67	2.67	17.4	110	4.35	47.2	94	16.5	300
Rb	37	1.08	0.81	18.8	108	4.65	50.9	93	16.8	314
Sr	38	1.79	2.08	12.5	108	3.12	52.2	93	11.1	318

Y	39	3.09	3.92	11.9	107	3.05	53.6	92	7.5	325
Zr	40	4.31	5.43	12.5	107	2.97	54.9	92	10.5	330
Nb	41	5.45	7.5	11.8	106	1.87	56.2	92	9.8	336
Mo	42	6.39	6.9	15.5	106	3.7	57.3	91	12.8	339
Tc	43	6.69	5	23.1	105	5.45	58.5	91	18.5	342
Ru	44	7.26	5.7	22.7	105	5.4	59.6	91	17.8	345
Rh	45	7.25	5.3	25.1	104	5.85	61.1	90	19.5	352
Pd	46	6.47	4.8	25.4	104	5.85	62	90	19.3	354
Ag	47	5.85	2.7	42.1	103	9.7	63.6	90	31.5	361
Cd	48	4.66	1.17	79.9	103	18	64.6	89	58.4	362
In	49	3.81	2.33	33.6	102	7.4	66	89	24.1	366
Sn	50	3.63	3	25.5	102	5	66.8	88	17.8	368
Sb	51	3.07	1.75	38	101	8.1	68.1	88	26.7	372
Te	52	2.94	2.41	27.1	101	5.7	68.5	87	17.9	372
Cs	55	0.85	8	2.55	99	1.38	73.1	87	1.59	389
Ba	56	1.66	2.2	12.9	99	3.87	74.3	87	11.6	391
La	57	2.67	4	16.7	99	3.42	75.2	83	10.2	397
Hf	72	4.48	7.3	21	93	3.8	96	83	9.6	469
Ta	73	5.52	8.7	22.2	93	4.1	97.7	83	10	472
W	74	6.09	8.76	24.7	93	4.5	98.7	83	11.1	476
Re	75	6.64	8.2	29.4	93	5.3	100.2	83	12.9	480
Os	76	7.07	7.5	34.7	92	6.2	101.4	82	14.8	484
Ir	77	6.99	7.1	36.9	92	6.6	102.8	82	15.6	488
Pt	78	6.61	5.56	45.4	92	8.1	104.3	82	18.7	496
Au	79	5.90	3.92	54.5	91	10.4	105.8	82	23.6	500
Hg	80	4.28	0.645	262	91	46	107.2	82	104	504
Tl	81	3.5	1.74	81	91	14	108.5	81	31.7	508
Pb	82	3.28	2.04	65	91	11.2	108.9	81	24.6	512
Bi	83	2.79	2.06	54.3	90	9.7	111.4	81	21.1	516
Th	90	3.09	6.5	21.9	88	3.68	121.3	80	7.2	548

TABLE 4.8. Sputtering Yields for Iron $^{26}Fe_{55.8}$ with Various Ions Calculated Using Smith's Formula [Eq. (4.27)]

Energy, eV	$^2H_4^+$	$^5B_{10}^+$	$^6C_{12}^+$	$^7N_{14}^+$	$^8O_{16}^+$	$^{18}Ar_{40}^+$	$^{36}Kr_{83.8}^+$
10	$1.3 \cdot 10^{-3}$	$3.6 \cdot 10^{-3}$	$4 \cdot 10^{-3}$	$4.6 \cdot 10^{-3}$	$5.3 \cdot 10^{-3}$	$1.3 \cdot 10^{-2}$	$2.8 \cdot 10^{-2}$
20	$2.6 \cdot 10^{-3}$	$7.1 \cdot 10^{-3}$	$7.8 \cdot 10^{-3}$	$9.2 \cdot 10^{-3}$	$1.05 \cdot 10^{-2}$	$2.6 \cdot 10^{-2}$	$5.5 \cdot 10^{-2}$
40	$5.1 \cdot 10^{-3}$	$1.4 \cdot 10^{-2}$	$1.6 \cdot 10^{-2}$	$1.8 \cdot 10^{-2}$	$2.1 \cdot 10^{-2}$	$5.3 \cdot 10^{-2}$	$1.1 \cdot 10^{-1}$
80	$9.95 \cdot 10^{-3}$	$2.8 \cdot 10^{-2}$	$3.1 \cdot 10^{-2}$	$3.6 \cdot 10^{-2}$	$4.2 \cdot 10^{-2}$	$1.05 \cdot 10^{-1}$	$2.2 \cdot 10^{-1}$
100	$1.2 \cdot 10^{-2}$	$3.5 \cdot 10^{-2}$	$3.9 \cdot 10^{-2}$	$4.5 \cdot 10^{-2}$	$5.2 \cdot 10^{-2}$	$1.3 \cdot 10^{-1}$	$2.8 \cdot 10^{-1}$
200	$2.3 \cdot 10^{-2}$	$6.7 \cdot 10^{-2}$	$7.5 \cdot 10^{-2}$	$8.9 \cdot 10^{-2}$	$1.0 \cdot 10^{-1}$	$2.6 \cdot 10^{-1}$	$5.5 \cdot 10^{-1}$
400	$4.0 \cdot 10^{-2}$	$1.3 \cdot 10^{-1}$	$1.4 \cdot 10^{-1}$	$1.7 \cdot 10^{-1}$	$2.0 \cdot 10^{-1}$	$5.1 \cdot 10^{-1}$	$1.1 \cdot 10^{0}$
800	$6.2 \cdot 10^{-2}$	$2.3 \cdot 10^{-1}$	$2.6 \cdot 10^{-1}$	$3.1 \cdot 10^{-1}$	$3.65 \cdot 10^{-1}$	$1.0 \cdot 10^{0}$	$2.1 \cdot 10^{0}$
1000	$6.9 \cdot 10^{-2}$	$2.7 \cdot 10^{-1}$	$3.1 \cdot 10^{-1}$	$3.75 \cdot 10^{-1}$	$4.4 \cdot 10^{-1}$	$1.2 \cdot 10^{0}$	$2.65 \cdot 10^{0}$
2000	$\underline{8.4 \cdot 10^{-2}}$	$4.2 \cdot 10^{-1}$	$5.0 \cdot 10^{-1}$	$6.2 \cdot 10^{-1}$	$7.4 \cdot 10^{-1}$	$2.2 \cdot 10^{0}$	$5.1 \cdot 10^{0}$
4000	$8.2 \cdot 10^{-2}$	$5.5 \cdot 10^{-1}$	$6.9 \cdot 10^{-1}$	$8.9 \cdot 10^{-1}$	$1.1 \cdot 10^{0}$	$3.85 \cdot 10^{0}$	$9.4 \cdot 10^{0}$
8000	$6.4 \cdot 10^{-2}$	$\underline{5.7 \cdot 10^{-1}}$	$\underline{7.7 \cdot 10^{-1}}$	$\underline{1.05 \cdot 10^{0}}$	$1.36 \cdot 10^{0}$	$5.9 \cdot 10^{0}$	$1.6 \cdot 10^{1}$
10 000	$5.6 \cdot 10^{-2}$	$5.5 \cdot 10^{-1}$	$7.6 \cdot 10^{-1}$	$\underline{1.05 \cdot 10^{0}}$	$\underline{1.37 \cdot 10^{0}}$	$6.5 \cdot 10^{0}$	$1.95 \cdot 10^{1}$
20 000	$3.5 \cdot 10^{-2}$	$4.3 \cdot 10^{-1}$	$6.2 \cdot 10^{-1}$	$9.05 \cdot 10^{-1}$	$1.24 \cdot 10^{0}$	$\underline{7.7 \cdot 10^{0}}$	$2.7 \cdot 10^{1}$
40 000	$2.0 \cdot 10^{-2}$	$2.8 \cdot 10^{-1}$	$4.2 \cdot 10^{-1}$	$6.35 \cdot 10^{-1}$	$9.0 \cdot 10^{-1}$	$7.2 \cdot 10^{0}$	$\underline{3.2 \cdot 10^{1}}$
80 000	$1.05 \cdot 10^{-2}$	$1.6 \cdot 10^{-1}$	$2.5 \cdot 10^{-1}$	$3.9 \cdot 10^{-1}$	$5.6 \cdot 10^{-1}$	$5.4 \cdot 10^{0}$	$3.0 \cdot 10^{1}$
100 000	$8.5 \cdot 10^{-3}$	$1.3 \cdot 10^{-1}$	$2.1 \cdot 10^{-1}$	$3.2 \cdot 10^{-1}$	$4.7 \cdot 10^{-1}$	$4.75 \cdot 10^{0}$	$2.8 \cdot 10^{1}$
200 000	$4.4 \cdot 10^{-3}$	$7.1 \cdot 10^{-2}$	$1.1 \cdot 10^{-1}$	$1.75 \cdot 10^{-1}$	$2.6 \cdot 10^{-1}$	$2.9 \cdot 10^{0}$	$2.0 \cdot 10^{1}$
400 000	$2.2 \cdot 10^{-3}$	$3.65 \cdot 10^{-2}$	$5.8 \cdot 10^{-2}$	$9.15 \cdot 10^{-2}$	$1.4 \cdot 10^{-1}$	$1.6 \cdot 10^{0}$	$1.2 \cdot 10^{1}$
800 000	$1.1 \cdot 10^{-3}$	$1.85 \cdot 10^{-2}$	$3.0 \cdot 10^{-2}$	$4.7 \cdot 10^{-2}$	$7.0 \cdot 10^{-2}$	$8.5 \cdot 10^{-1}$	$6.8 \cdot 10^{0}$
1 000 000	$8.9 \cdot 10^{-4}$	$1.5 \cdot 10^{-2}$	$2.4 \cdot 10^{-2}$	$3.8 \cdot 10^{-2}$	$5.6 \cdot 10^{-2}$	$6.9 \cdot 10^{-1}$	$5.5 \cdot 10^{0}$

Note. The maximum values of the sputtering yields have been underlined.

This formula makes it easy to obtain estimates of the sputtering yield, in particular for iron (Table 4.8). Comparing the computed values with experiment (see Fig. 4.5) for Ar^+ ions, we see that Eq. (4.27) gives a maximum sputtering yield for iron of 6.6, while the experimentally measured value is 2 in the energy range 6-22 keV. The discrepancy is still greater for the heavier Kr^+ ions. Equation (4.27) gives a sputtering yield of 27 at 20 keV, while the experimental value is only 4, and Eq. (4.27) gives a maximum of 32 at 40 keV. Other experimental data for pure iron [77] irradiated by 10-20 keV O_2^+ ions give a sputtering yield of 0.6, for N_2^+ ions the ratio is 1.1-1.25, and for Ar^+, 1-2.5. The discrepancy between the sputtering yield given by Eq. (4.27) and the experimental values becomes significant with Ne^+ ions and reaches one or two orders of magnitude with Ar^+, Kr^+, and Xe^+ ions, for example, during sputtering of Be (Fig. 4.6). Equation (4.27), therefore, can only be used for crude estimates.

In order to reduce the discrepancy between the calculated and experimental values of the sputtering yield, particularly at low energies, Smith [78] has refined Eq. (4.27):

$$S = \frac{20}{U_0} Z_1^2 Z_2^2 \frac{M_1}{M_2} \frac{E - E_t}{(E - E_t + 50 Z_1 Z_2)^2} . \qquad (4.27a)$$

This correction, however, is negligible for ion energies that exceed the threshold energy by tens, hundreds, or thousands of times. It is necessary to introduce other correction factors, which will primarily replace the constant factor 20 in Eq. (4.27a), for each ion–target atom pair on the basis of experimental data.

A rigorous statistical examination of the sputtering process, including detailed angular and energy characteristics of the sputtering cascades and the effect of the target boundary [79, 80], has made it possible to obtain analytic expressions for the angular distributions and energy spectra of sputtered atoms and for the sputtering yields of single element and binary alloys bombarded by light ions (H^+, D^+, He^+) at energies of 10^2-10^5 eV. Simple formulas without fitted parameters have been obtained:

a) for $Z_1 < Z_2$

$$Y = 0.4 \frac{\Sigma_i}{\Sigma_i + \Sigma_a} \left[\frac{E_0}{2U_i} - \frac{0.3 U_i}{E_0} - 1 + \frac{0.8 U_i^2}{E_0^2} \left(1 + \ln \frac{E_0}{U_i} \right) \right]; \qquad (4.28)$$

b) for $Z_1 > Z_2$

$$Y = 0.4 \frac{\Sigma_i}{\Sigma_i + \Sigma_a} \left[\frac{E_0}{2 U_i} + \frac{U_i}{E_0} \left(3.5 + 3 \ln^2 \frac{E_0}{U_i} \right) - 4 \right], \qquad (4.29)$$

where Σ_i and Σ_a are the total cross sections for collisions of the ion with atoms and of knock-on atoms with target atoms; E_0 is the average energy of the primary knock-on atoms; and U_i is the surface binding energy for each type of atom in a multicomponent material. For multicomponent mixtures U_i can be estimated from the diffusion activation energy Q ($Q \approx 0.6 U_0$).

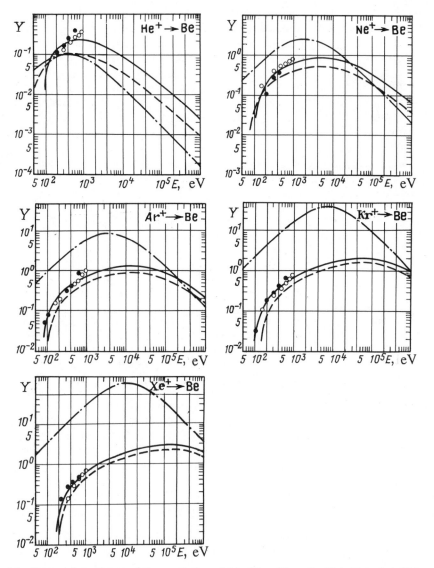

Fig. 4.6. Energy dependence of the sputtering yield of beryllium for He⁺, Ne⁺, Ar⁺, Kr⁺, and Xe⁺ ions according to Eq. (4.27) (–·–·–) and to the formula of Matsunami (– – –) [61]; • and O denote experimental data [67].

The surface energy of two-component materials is given roughly in terms of the binding energy of the surface atoms in the pure materials, U_{10} and U_{20}, and their concentrations in the alloy, p_1 and p_2, by

$$U_i = (p_1 U_{10} + p_2 U_{20}) (p_2 + 1)^{-1}.$$

(4.29a)

TABLE 4.9. Comparison of Experimentally Measured (Y_{meas}) and Calculated (Y_{calc}) Values of the Sputtering Yield of Silver Bombarded by Atomic and Molecular Ions of Antimony [82]; h is the Radius of a Hemispherical Crater Containing Y_{meas} Atoms and $\int_0^h F_D(x)dx$ is the Ion Energy Released over a Path Length h

Ion	Energy, keV	Y_{meas}	Y_{calc}	h, nm	$\int_0^h F_D(x)\,dx,$ eV
Sb^+	10	15	13,4	0.496	402
	30	25	17,5	0.588	561
	70	37	18.9	0.67	657
Sb_2^+	10	54	22,2	0.759	1197
	30	101	29.2	0.936	1809
	90	160	36.6	1.09	2295
Sb_3^+	21	184	36.1	1,145	3642
	45	301	43.8	1.349	4452
	90	808	47.7	1.874	7236
Bi^+	30	46	21.7	0.721	927
	100	82	25.5	0.874	1149
	200	88	25.3	0.895	1164
Bi_2^+	60	365	43.4	1,439	4251
	90	518	47.8	1.616	5208

Note: The values of Y corresponding to nonlinear sputtering are underlined.

The sputtering characteristics of a two-component Fe–C target (0.8 Fe, 0.2 C) bombarded by 1000-eV He^+ ions along the normal to the target surface have been calculated. The energy spectra and angular distributions of sputtered atoms have been obtained, as well as the selective sputtering yields $Y_{Fe} = 0.137$ and $Y_C = 0.082$. For pure iron $Y_{Fe} = 0.154$, which is in agreement with the experimentally measured 0.15 [66]. In the calculations the values $U_{Fe} = 4.3$ (energy of sublimation) and $U_C = 2.2$ eV were used.

The accuracy of the calculated values of the sputtering yield is estimated as no better than 50%. It depends upon the accuracy with which the Linhard formula and the Born–Mayer potential describe the ion–atom and atom–atom collisions.

4.6. Nonlinear Effects in Ion Sputtering

In a number of experimental papers [81-88] it has been found that the sputtering yield for heavy, particularly molecular, ions with energies of 50-100 keV and above are greater than the values calculated according to Sigmund's theory. Measurements of the sputtering yield for gold (in the form of 0.5-μm-thick polycrystalline layers deposited on quartz) bombarded by 18 different ions at energies of 45 keV showed [81] that for $Z_1 = 40$-80 the yield exceeds the calculations by a factor of 2-5. For Ar^+, Al^+, Zn^+, and, more noticeably, Tm^+ ions, the sputtering yield was found to decrease with the implanted ion dose. For example, for sputtering by Ar^+ ions the yield decreased from 17.5 to 11.5 after removal of a 170-nm-thick layer of gold foil. For Kr^+, Xe^+, and Pb^+ at low doses the sputtering yields were 37, 53, and 82, respectively, and after large doses they fell to 25, 35, and 55.

TABLE 4.10. Average $(\overline{Y})$ and Maximum (Y_{max}) Sputtering Yields of Gold for Monatomic and Diatomic Bismuth Ions [85]

Parameter	Bi^+			Bi_2^+		
Ion energy, keV	50	80	125	25	50	125
$\overline{Y}$	1200	2200	1900	2500	2800	3800
Y_{max}	4000	9500	13000	6500	20000	30000
E_D/Y_{max}, eV/atom	9.4	6.2	6.9	5.9	3.8	6.0

During sputtering of gold by 150-keV Se^+ ions the sputtering yield is roughly 33, while for 300-keV Se_2^+ the yield is 40-46. The sputtering yields of gold and silver were observed to increase by 44% during irradiation by Se_2^+. When these materials are irradiated by Te_2^+ ions, the sputtering yields increase by factors of 2.15 and 1.67, respectively, compared to irradiation with Te^+.

Thomson [82] has also done experiments with thin polycrystalline films (30-100 nm) of silver, gold, and platinum on silicon and silicon dioxide. The dependence of the sputtering yield for these metals on the energy of Sb^+, Sb_2^+, Sb_3^+, Bi^+, and Bi_2^+ ions has been studied in the range 10-250 keV. Some results of these experiments are shown and compared with the sputtering yields calculated using Sigmund's formula in Table 4.9. The experimental values are considerably higher than those calculated with Eq. (4.6), especially for the di- and triatomic molecules.

Nonlinear sputtering of silver and gold sets in when the energy release approaches 10-20 eV·nm^2 per atom. A further increase in the specific energy release makes the sputtering yield proportional to the cube of this quantity, i.e.,

$$Y \sim F_D(0)^3.$$

Another explanation for the nonlinear dependence of the sputtering yield is a reduction in the effective binding energy, for example from 3.04 to 0.2 eV for silver. There may be an extra contribution to the atom yield from ions that are not incident along the normal to the surface of a crater.

The sputtering of UF_4 targets by ^{16}O, ^{19}F, and ^{35}Cl ions with energies of 0.12-1.5 MeV/amu and of frozen SO_2 ice by He^+ (1.5 MeV) and F^+ ions (1.5-25 MeV) has been studied [83, 84]. In the first paper, sputter coated layers of $\approx$200 nm thick UF_4 (enriched to 93% ^{235}U) on thick copper or carbon substrates were used. During ion bombardment a maximum in the sputtering yield was observed near energies of 0.34-0.7 MeV/amu, which was 7, 24, and 200-450, respectively, for the ions of ^{16}O, ^{19}F, and ^{35}Cl. In this energy interval the sputtering yield was proportional to the fourth power of the specific energy release, i.e.,

$$Y \sim [dE/dx]^4.$$

An electron microscope has been used for transillumination of thin (60-80 nm) gold films irradiated by 10-500 keV Bi^+ and Bi_2^+ ions to observe the formation of craters with diameters of about 5 nm accompanied by extremely large sputtering yields [85] (Table 4.10). The breakup energy E_D (equal to the energy of the cascade minus the losses in exciting electrons in the cascade) given in Table 4.10 is greater than the sublimation energy of gold atoms (3.8 eV). With monatomic bismuth ions the threshold for formation of visible craters is around 50 keV/atom and with diatomic ions, 12 keV/atom. The number of visible craters per Bi^+ ion increases from below 10^{-3} to roughly 10^{-2} in the energy range 50-125 keV. The crater sizes range from 1 to 13 nm. At an ion energy of 125 keV/atom the average diameters of the craters formed by monatomic and diatomic bombardment are 4.6 and 5.4 nm, respectively. Extremely high sputtering yields (up to $3 \cdot 10^4$) are explained [85] by statistical fluctuations in the energy release in collision cascades.

Pramanik and Seidman [86] have used an autoion microscope to observe nonlinear effects. At a level of single atom resolution they observed damage to a tungsten needle with a radius of curvature of about 35 nm by 20-keV Ag^+ and W^+ ions and 40-keV Ag_2^+ and W_2^+ ions at ion doses of $6 \cdot 10^{12}$ and $3 \cdot 10^{12}$ cm^{-2}, respectively. The number of vacancies produced per dimer ion (of Ag) was 1.55 times that created per monomer, while the tungsten ions did not show such a contrast. The average size of the depleted zone created by Ag^+ and W^+ ions was 1.75 nm and by Ag_2^+ and W_2^+ ions, 2.85 nm. The depth of the zones depleted of tungsten atoms was 1-6 nm in the first case, and was 0.5-3.6 nm for irradiation by molecular ions. The corresponding number of vacancies in this zone varied over the ranges 85-200 for the monomers and 190-675 for the dimers.

Baranov et al. [87] have reviewed the experimental data on the sputtering of materials by nuclear fission fragments. Their analysis of these data and their own studies and theoretical calculations led them to the following conclusions. At low energies (≤ 10 keV/nucleon), when the heavy ions are not multiply charged, every material is sputtered as a result of elastic collisions. The sputtering yield can be estimated using Sigmund's cascade theory. When the specific losses by heavy multiply charged ions through ionization are on the order of $1.3 \cdot 10^4$ eV/nm, two types of sputtering apparently begin to occur. The first is characteristic of fine-grained samples containing grains with sizes of up to 20 nm in the surface layer. In the case of nonconducting materials the thickness of the layers may be arbitrary. Large values of the sputtering yield ($5 \cdot 10^2 - 5 \cdot 10^4$) are typical for this type of sputtering. However, the yield drops sharply (to about 30) if the diameter of the grains in the layers increases from 5 to 20 nm. The thickness of the layer of material from which sputtering takes place is 3-7 nm, which is close to the grain size. Fine-grained layers sputter either as individual atoms or as relatively small groups of atoms, but often in the form of small pieces with dimensions of about 0.1-1 μm or more. Another type of sputtering is characteristic for coarse-grained samples ($\varphi > 50$ nm). Relatively small sputtering yields (5-50) are typical of this type of sputtering.

The first type of sputtering is represented by a model of an isolated grain which, depending on its size and other parameters, either undergoes complete sputtering or does not sputter at all. An ion explosion or thermal electron peak is responsible for the sputtering. A formula has been proposed for the sputtering yield [87]:

$$Y = (4/3)\pi^2 N \sum_i R_i^5 f_i (1 - 1/n_i^2)/\Pi.$$ (4.30)

Here N is the number of atoms per unit volume, $N = 6.025 \cdot 10^{23} \rho/M_2$ cm^{-3}; f_i is the number of grains of radius R_i on the irradiated surface Π; $(1 - 1/n_i^2) = P_i$ is the sputtering probability for a grain with radius R_i; $6.02 \cdot 10^{23}$ is the Avogadro number (g-mole)$^{-1}$; and

$$n_i = (dE/dx)_e/(dE/dx)_{e \text{ min}, i},$$ (4.31)

where $(dE/dx)_{e \text{ min},i}$ is the minimum value of the ionization slowing-down power, at which a heavy multiply charged ion, passing along the diameter of a grain, releases enough energy to sputter a grain with $d < 20$ nm into separate atoms. A dose dependence has been noted for the first type of sputtering. The sputtering yield for integrated fluxes of nuclear fission fragments in the range 10^{13}-10^{15} cm^{-2} decreases to 15-30 atoms per fragment, possibly as a result of coagulation of grains to roughly 20 nm during irradiation by fission products.

Sputtering by the action of a shock wave has been proposed as an explanation of nonlinear sputtering [88]. When the average velocity of the atoms in a cascade exceeds the speed of sound in the target material, a front of atoms may serve as a "hammer" which compresses the surrounding medium in the backward direction. Upon reaching the free surface, the shock wave is reflected. When the pressure on this surface exceeds a critical value, some portion of the target is broken off from the surface. An expression has been obtained for the sputtering yield

$$Y \approx \frac{\pi \rho_0}{3} \left[\frac{3 S_n(E)}{2\pi \, \epsilon_c \, \rho_0} \right]^{3/2} (\sec^2 \theta - 1),$$ (4.32)

where ρ_0 is the density of target atoms under normal conditions; ϵ_c is the energy density of atoms in the volume of a cascade with radius R_c, i.e.,

$$\epsilon_c = \frac{3 E_D}{4\pi R_c^3 \rho_0} = \frac{3 \int_0^{2R_c} S_n(E) \, dz}{4\pi R_c^3 \rho_0};$$ (4.33)

$S_n(E)$ is given by Eqs. (4.8) and (4.10), and θ is the half angle of the cone broken off from the material by the shock wave.

In the energy range 10^0-10^4 keV, choosing $\epsilon_c = 16$ eV for copper gives good agreement between this formula and experimental data for sputtering by Ar$^+$, Cu$^+$, Xe$^+$, Kr$^+$, and Hg$^+$ ions and for silver, with $\epsilon_c = 17$ eV, by Kr$^+$, Ag$^+$, Xe$^+$, and Hg$^+$ ions. Nonlinear sputtering sets in for specific energy losses of 1000-3000 eV/nm in nuclear collisions.

The dependence of the average diameter of the crater formed in silver by irradiation with Kr^+, Ag^+, Xe^+, and Hg^+ ions in the range 10^0-10^4 keV has been computed. It varies from 0.7-1.1 to 2.0-2.3 nm.

The sputtering of solids into clusters (polyatomic structures), such as Al_n^+ ($n = 1$-17), V_n^+ ($n = 1$-16), Cu_n^+ ($n = 1$-39), Ag_n^+ ($n = 1$-41), Au_n^+ ($n = 1$-21), Nb_n^+ ($n = 1$-17), Ta_n^+ ($n = 1$-13), Mo_n^+ ($n = 1$-17), W_n^+ ($n = 1$-17), has been observed by Dzhemilev et al. [89]. In order to explain sputtering into clusters Bitenskii and Parilis [89] have proposed a cascade-hydrodynamic mechanism and sputtering theory. When cones are present on the surface, the development of a high-density collision cascade and the propagation of a shock wave to the vertex of a cone leads to accumulation of energy, and the chipping off and ejection of the top of the cone in the form of a large cluster. Clusters are primarily emitted in directions close to the surface normal. Experiments show that the yield of large clusters decreases with the number of particles according to a power law.

4.7. Dependence of the Sputtering Yield on the Angle of Incidence of an Ion at the Target Surface

For ions with medium and higher masses the sputtering yield increases as the angle of incidence (relative to the normal to the surface) of the ions is raised, reaching a maximum at 60-70° (Fig. 4.7). The relative variation in the sputtering yield is greater for higher ion energies. The sputtering yields for 1-8 keV light ions (H^+, D^+, He^+) on a nickel target increased all the way to 80°, reaching values that were 4.5-11 times greater than at normal incidence. At an angle of 80° the yields for sputtering of tungsten and molybdenum by 4- and 8-keV protons were 15.5 and 20 times, respectively, greater than at normal incidence [66].

Sigmund's theory gives the following dependence of the sputtering yield on the angle of incidence:

$$Y(\theta) = Y(0)/\cos^f\theta. \qquad (4.34)$$

The exponent $f = 1.7$-1.0 for $M_2/M_1 = 0.1$-10. For $M_2/M_1 = 1$, $Y(\approx 90°) = (4.5$-$6.0)Y(0)$. For $M_2/M_1 = 0.1$-0.6, the theory does not give a unique answer. The approximation

$$Y(\theta) = Y(0) \exp[-0.1 \lambda N^{1/3} (1 - \cos\theta)] \, \alpha'(\theta)/\alpha'(0), \qquad (4.35)$$

where $\alpha'(\theta)/\alpha'(0)$ is the ratio of the energy accomodation coefficients for oblique and normal incidence of the ions, has been obtained by Ryzhov and Shkarban [75]. For $\theta \leqslant 45°$, $\alpha'(\theta)/\alpha'(0) \approx 1$; $Y(0)$ is given by Eq. (4.23).

Firsov [90] has obtained a formula which gives the optimum grazing angle for an ion (Table 4.11) at which the sputtering yield reaches its maximum

$$\xi_0 \approx \sqrt{\frac{h_{eff}^2 \, N^{2/3}}{mE}} \, \frac{Z_1 Z_2}{(\sqrt{Z_1} + \sqrt{Z_2})^{2/3}} \left[0.63 \, h_{eff} - \frac{40 b_0 \, N^{1/3}}{(\sqrt{Z_1} + \sqrt{Z_2})^{2/3}} \right], \qquad (4.36)$$

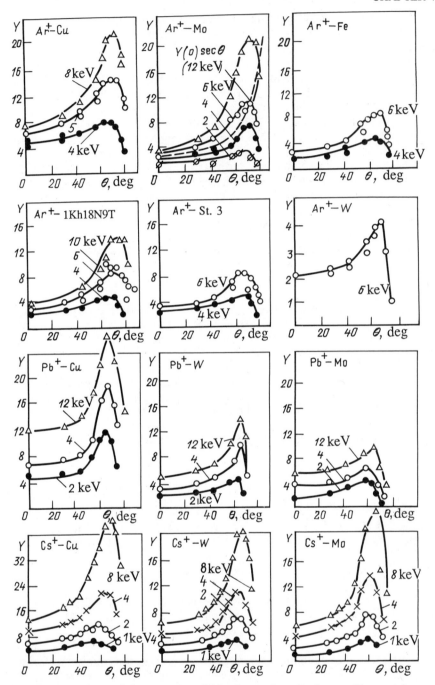

Fig. 4.7. Dependence of the sputtering yields of a number of polycrystalline metals on the angle of incidence of ions at different energies [75].

TABLE 4.11. Comparison of the Theoretical Optimum (ξ_{to}) and Experimental Optimum (ξ_{eo}) Grazing Angles and the Effective Depth Responsible for Sputtering [90]

Target ion	Particle energy, keV	ξ_{to}, deg	ξ_{eo}, deg	h_{eff} $N^{-1/3}$	h_{eff} nm
Cu – N$^+$	28	7	5	0.87	0.20
Cu – N$^+$	14	10	12	9.2	2.1
Cu – Ne$^+$	28	8	8	5.0	1.14
Cu – Ne$^+$	14	12	12	5.4	1.23
Cu – Ar$^+$	28	10	13	9.4	2.15
Cu – Ar$^+$	27	10	12	7.5	1.71
Cu – Ar$^+$	27	10	10	4.2	0.95
Cu – Ar$^+$	8	18	14	1.1	0.25
Cu – Ar$^+$	4	25	24	3.2	0.73
Cu – Cs$^+$	8	27	17	0.43	0.098
Mo – Ar$^+$	27	9	10	4.0	0.465
Mo – Ar$^+$	12	13	12	2.0	0.23
Cu – Xe$^+$	18	16	20	1.1	0.128
Ag – Ne$^+$	30	7	7	2.9	0.61
Ag – Ar$^+$	30	9	9	3.0	0.63
Ag – Kr$^+$	25	12	12	2.3	0.48
Ni – Ar$^+$	27	10	10	3.5	0.68
Fe – Ar$^+$	6	16	15	0.85	0.19
Fe – Ar$^+$	4	19	20	1.65	0.375

where $h_{eff} \sim (\Delta E/E_d)^{1/3}$ is the effective depth responsible for sputtering; ΔE is the mean energy lost by a bombarding particle over the path length $N^{-1/3}$; E_d is the energy for an atom to escape from the surface; N is the number of atoms per unit volume; m is the electron mass; and $b_0^2 = \hbar^2/(me^2)$ is the Bohr radius. As can be seen from Table 4.11, Eq. (4.36) is in good agreement with, and in 37% of the cases, is absolutely the same as, the experimental data. The theory and experiments discussed by Firsov [90] indicate that for ions of medium mass the sputtering yield has a maximum at angles of incidence of 63-85°.

Sputtering yields for processes involving two mechanisms have been examined [91]:

$$Y(E, \theta) = Y_I (E, \theta) + Y_{II} (E, \theta). \tag{4.37}$$

Mechanism I causes sputtering during collision of an ion with an atom lying on the surface and mechanism II, during reflection of an ion from lower-lying layers of atoms. The binary collision model gives

$$Y_{II} (\theta) = Y_{II} (0)/\cos \theta. \tag{4.38}$$

Mechanism I gives a stronger dependence on the angle of incidence of the ions.

Some quantitative data on the dependence of the sputtering yield for nickel on the angle of incidence of the ions are given in Table 4.12.

TABLE 4.12. Dependence of the Sputtering Yield of Nickel on the Angle of Incidence θ of H$^+$ and He$^+$ Ions with Different Energies E_1 [91]

θ, deg	For H$^+$ with E_1, keV			For He$^+$ with E_1, keV
	0,45	1	4	4
0	$9.82 \cdot 10^{-3}$	$(1.55-1.69) \cdot 10^{-2}$	$1.36 \cdot 10^{-2}$	$(1.8-1.97) \cdot 10^{-1}$
20	–	$2.23 \cdot 10^{-2}$	$1.7 \cdot 10^{-2}$	$2.33 \cdot 10^{-1}$
40	$1.5 \cdot 10^{-2}$	$3.17 \cdot 10^{-2}$	$2.83 \cdot 10^{-2}$	$3.07 \cdot 10^{-1}$
60	$1.96 \cdot 10^{-2}$	$(4.56-5.2) \cdot 10^{-2}$	$4.58 \cdot 10^{-2}$	$(3.62-4.52) \cdot 10^{-1}$
70	–	$(6.8-7.58) \cdot 10^{-2}$	$8.6 \cdot 10^{-2}$	$(5.08-6.41) \cdot 10^{-1}$
80	$\underline{2.2 \cdot 10^{-2}}$	$\underline{1.03 \cdot 10^{-1}}$	$\underline{1.43 \cdot 10^{-1}}$	$\underline{8.22 \cdot 10^{-1}}$
85	–	$6.35 \cdot 10^{-2}$	–	$7.69 \cdot 10^{-1}$

θ, deg	For H$^+$ with E_1, keV	For He$^+$ with E_1, keV	For He$^+$ with E_1, keV
	50	100	50
0	$4.9 \cdot 10^{-3}$	$(4.6-6.4) \cdot 10^{-2}$	$5.9 \cdot 10^{-2}$
30	$6.6 \cdot 10^{-3}$	$7.1 \cdot 10^{-2}$	–
40	–	–	–
60	$1.45 \cdot 10^{-2}$	$1.5 \cdot 10^{-1}$	–
80	$4.5 \cdot 10^{-2}$	$3.0 \cdot 10^{-1}$	–

Note: The maximum values of Y have been underlined.

Yamamura, Itikawa, and Itoh [92] have analyzed in detail the results of 35 papers on the angular dependence of the sputtering yield and have proposed the following empirical formula:

$$\frac{Y(\theta)}{Y(0)} = t^f \exp\left[-\Sigma(t-1)\right], \tag{4.39}$$

where θ is the angle of incidence relative to the surface normal; $t = 1/\cos\theta$; and the parameters f and Σ are determined by the method of least squares.

$$\Sigma = f \cos\theta_{opt}, \tag{4.40}$$

where θ_{opt} is the angle of incidence at which the maximum yield is observed. For light ions they propose the formula

$$Y(0) = 0.042 \frac{F_D(E^*) R_N(E)}{U_0} \left[1 - \left(\frac{E_{th}}{E}\right)^{1/2}\right]^{2.8}, \tag{4.41}$$

where $F_D(E^*)$ is the energy released in the surface layer of a solid by a reflected ion; E^* is the average energy of a reflected ion; $R_N(E)$ is the particle reflection coefficient; and E_{th} is the threshold energy for normal incidence. For heavy ions,

$$\frac{Y(\theta)}{Y(0)} = (\cos\theta)^{-f_s} \left[\frac{1 - (E_{th}/E)^{1/2}\cos\theta}{1 - (E_{th}/E)^{1/2}}\right]. \tag{4.42}$$

TABLE 4.13. Ratio of the Maximum Sputtering Yield to the Sputtering Yield for Normal Incidence, $Y_{max}(\theta)/Y(0°)$, and the Optimum Angle of Incidence θ_{opt}

Ion	E_1, keV	Target	$Y_{max}(\theta)/Y(0°)$	θ_{opt}, deg
He+	100	Ti	28	86
H+	4	Ni	13–18.5	82–84
H+	8	Ni	15	85
H+	50	Ni	32	87
H+	4	Ni	9.5	83
H+	8	Ni	12,7	84
He+	4	Ni	4.5–5	78–79
He+	8	Ni	5,8	79
He+	100	Ni	16,7	85
H+	50	Cu	50	88
H+	2	Mo	18	85
H+	4	Mo	26	85
H+	8	Mo	35	85
H+	50	Mo	86	87
D+	2	Mo	16	83
D+	8	Mo	33	85
D+	100	Mo	122	87
He+	4	Mo	9	80
He+	8	Mo	12,8	81
He+	50	Mo	33	85
He+	100	Mo	47	85
Ar+	150	Ti	9,7	85
Ar+	900	Ti	17,5	85
Hg+	30	Fe	4,8	76,5
Hg+	0,4	Fe	12	59
Hg+	0,8	Fe	7	62
Ar+	1.05	Ni	2	66
Hg+	0,2	Ni	2,5	50
Hg+	0,8	Ni	2,3	52–61
Ar+	1.05	Cu	1.7–3	66–72
Ar+	27	Cu	3,5–4,5	79–81
Ar+	37	Cu	2,–4,8	79
Ar+	100	Cu	6,5	83
Kr+	1.05	Cu	2	65–69
Kr+	45	Cu	5,5	79
Xe+	0,55	Cu	2–3,5	62
Xe+	1,05	Cu	2.4–2,8	62
Xe+	9.5	Cu	2.8–3,5	69–74
Xe+	30	Cu	3,5–4,8	78–81
Xe+	30	Mo	5–6	77–78
Hg+	20	Mo	4,2	76,5
D+	12.2	Nb	60	86
Nb+	60	Nb	6,7	81
Xe+	9,5	W	2–3.3	73–77
Xe+	30	W	3.3–4.4	77
Hg+	0.4	W	2–3	54–57
Hg+	0.8	W	2–3	57–61

Note: Conditions under which the increase in Y is extremely large have been underlined.

Table 4.13 shows calculated and experimental values of the ratio of the sputtering yield for optimum and normal incidence [91].

4.8. Dependence of the Sputtering Yield on the Ion Dose and the Residual Oxygen and Gas Pressures

The sputtering yield of metals coated with oxides is usually lower than for pure metals. Thus, a certain ion dose is required to sputter the oxide layer. But even a pure metal surface is oxidized by residual oxygen at the pressure in a vacuum vessel, $1.3 \cdot 10^{-1}$-$1.3 \cdot 10^{-4}$ Pa. Each ion–metal combination requires a definite current density at a given ion energy for contaminants to be automatically removed from the surface [62]:

$$j_+ \ Y_{film} \gg f \, j_g, \tag{4.43}$$

where j_+ and j_g, respectively, are the current densities of ions and atoms of the residual gas to the irradiated surface; Y_{film} is the sputtering yield for the film that contaminates the surface; and f is the attachment coefficient for molecules and atoms of the residual gas. It is known that

$$j_g = 2.6 \cdot 10^{20} \, pf \, (1 - \theta) / \sqrt{M_0 T}, \tag{4.44}$$

where $\theta = n/n_1$ is the degree of covering; n and n_1 are the number of adsorbed atoms and substrate atoms per cm^2 of the surface; and p is the gas pressure (Pa). For air,

$$j_+ \geqslant 0.45 \ \frac{p}{Y_{film}} \ f(1 - \theta). \tag{4.45}$$

The time (s) required to form a monomolecular layer of residual gas atoms is roughly $1/(7.5 \cdot 10^3 \, p)$. At a pressure of $1.3 \cdot 10^{-4}$ Pa a single monomolecular layer can form on a metallic surface in 1 s. In order to remove it when $Y_{film} = 1$ and $f = 1$, an argon ion current density on the order of 1 mA/cm^2 is needed.

Even for pure surfaces, such as gold, a dose dependence of the sputtering yield is observed [93]. During bombardment by 600-eV N_2^+ ions the sputtering yield for gold (measured without taking the ion–electron emission coefficient into account) initially increases rapidly from 0.1 to 0.7 as the ion dose reaches $5 \cdot 10^{15}$ cm^{-2}. The flux density of atoms removed from the surface (per cm^2) is $2.25 \cdot 10^{15}$ cm^{-2} (roughly 1.5 monolayers of gold). The steady-state value $Y/(1 + \gamma) = 0.85$-1 is reached after an ion dose of $5 \cdot 10^{16}$ cm^{-2}, which corresponds to the removal of 17-20 layers of gold ($\gamma = 0.15$ is the ion–electron emission coefficient). After air is let into the vessel, the sputtering yield again begins to increase rapidly, reaching the steady- state value. This is evidently related to the removal of an adsorbed layer from the surface.

Laser fluorescence spectroscopy has been used to measure the sputtering yield of polycrystalline iron (99.99%) bombarded by 2-keV H^+ and D^+ ions and by

6-keV He^+ ions and to examine the effect of the ion dose and of the ratio of the oxygen pressure to the current density [94]. For He^+, D^+, and H^+ ion doses of up to $1 \cdot 10^{17}$, $3 \cdot 10^{17}$, and $1 \cdot 10^{18}$ cm^{-2}, the sputtering yields are correspondingly low: $7 \cdot 10^{-3}$, $3 \cdot 10^{-3}$, and $1 \cdot 10^{-3}$. Increasing the He^+ dose further to $4 \cdot 10^{17}$ cm^{-2} raised the yield to $39 \cdot 10^{-3}$, but still without bringing it to the steady-state value. Increasing the D^+ dose to $3 \cdot 10^{18}$ cm^{-2} led to a stationary value of $(35 \pm 5) \cdot 10^{-3}$ after a rapid rise, and increasing the H^+ dose to $6 \cdot 10^{18}$ cm^{-2} led to a stationary value of $(16 \pm 1) \cdot 10^{-3}$.

When oxygen was fed in up to pressures corresponding to a small ratio of the rate of molecular influx to the rate of ion bombardment, the yield of iron atoms for new targets was initially small but increased by a factor of 10-15 after roughly 20 monolayers of iron had been removed. This increase correlated with the appearance of a rough surface structure. At higher values of the ratio of the flux of oxygen molecules to the ion flux (above 2-10), the surface corresponded to a continuously renewing oxidized surface. The sputtering yield remained low, but the original rough surface was smoothed. In this case, however, oxygen molecules were sputtered with a yield of about one O_2 molecule per incident ion. A change in the topography of the surface and in the binding energy of the atoms on the surface cannot explain the factor of 10-15 difference in the sputtering yields.

Under typical bombardment conditions the density of incident particles is roughly equal to the density of the surface atom layer. This leads, first, to the formation of a large number of displaced atoms [about $E_1/(2E_d)$ per incident ion]. Depending on the target temperature, they may either be burnt off or form an amorphous layer. Second, because of sputtering the surface layer moves with a speed given by Eq. (4.1) (up to tens of atomic layers/s). Third, the bombarding ions penetrate to a depth determined by their mean free path R in the solid. Thus, the properties of the surface layer are changing from the beginning. The time to reach an equilibrium state (relaxation time) can be estimated as [95]

$$\tau = R/v_s. \tag{4.46}$$

Depending on the conditions for the ion irradiation, the relaxation time may be minutes or seconds.

When the product of the residual gas pressure and the distance between the sputtering target and vessel walls or collectors is large ($pd \geq$ 10-40 Pa·cm), the bulk of the sputtered atoms is returned to the target because of collisions with residual gas atoms. The "apparent" sputtering yield (Y_{app}) is lower than the actual yield, and it can be found using the formula of Penning and Moubis [62]

$$Y = Y_{app} \frac{(2.3 \lambda_1 + p_0 d)}{2.3 \lambda_1}, \tag{4.47}$$

where λ_1 is the mean free path of an evaporated atom in the gas at a pressure of 133 Pa and p_0 is the gas pressure at 0°C, 133 Pa.

TABLE 4.14. Change in the Apparent Sputtering Yield as the Product $p_0 d$ (for $d = 1$ cm) Is Varied. Calculations Have Been Done for Argon and Air Using Eq. (4.47)

$p_0 d$, Pa·cm	$1.3 \cdot 10^{-2}$	$1.3 \cdot 10^{-1}$	1.3	13	133
Y_{app} / Y	0.99	0.96	0.55	0.10	0.01

TABLE 4.15. Dependence of the Sputtering Yield on the Different Cleavage Planes of Monocrystalline Copper Irradiated by 5-keV Argon Ions [62] and of Pure Iron Irradiated by 20-keV O_2^+ Ions [77], and on the Texture of Tungsten Layers on a Molybdenum Pipe for 4.5-keV Argon Ions [96]

Copper monocrystal

Cleavage plane	(110)	(100)	(023)	(012)	(122)	(113)	(111)
Y	2.6–2.9	4.2	4.9	5.6	6.8	7.2–7.5	9.3–9.8

Iron monocrystal

Cleavage plane	(110)	(100)	(111)
Y	0.68	0.62	0.55
Surface binding energy of atoms, eV	5.48	5.47	4.72

Texture of tungsten

Plane of sputtered layers	(110)	(100)	(111)
Y	0.67	0.92	1.15
Average rate of sputtering, μm/h	1.73	2.40	3.00
Surface binding energy of atoms, eV	11.86	11.52	9.75

The magnitude of the effect of reverse diffusion on the "apparent" sputtering yield can be seen from Table 4.14.

4.9. Dependence of Sputtering Yield on the Cleavage Plane of Monocrystals and on the Texture, Roughness, and Porosity of Target Surfaces

The sputtering yield depends strongly on the cleavage plane of monocrystals and on the texture of metal surfaces (Table 4.15).

The sequence of sputtering yields for the different planes of textured tungsten corresponds to the minimum binding energies required to move an atom to infinity given by Jackson [153].

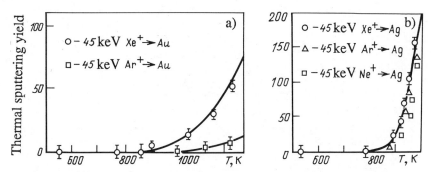

Fig. 4.8. Dependence of the thermal sputtering yield on the temperature of gold (a) and silver (b) samples during bombardment by 45-keV Xe^+, Ar^+, and Ne^+ ions [101].

TABLE 4.16. Dependence of the Sputtering Yield of Iron and Nickel on the Energy of Cs^+ Ions and on the Target Temperature

Metal	T, °C	Sputtering yield for E_1, keV								
		2	3	4	5	6	7	8	9	10
	700	2.10	2.64	3.0	3.48	3.86	4.10	4.21	4.17	4.21
	800	2.33	2.83	3.17	3.62	4.34	4.26	4.26	4.26	4.26
Fe	900	2.50	3.13	3.61	4.00	4.16	4.78	4.86	4.84	4.89
	1000	2.63	3.61	4.25	4.80	5.24	5.52	5.63	5.56	5.59
	1100	2.79	3.59	4.31	4.90	5.40	5.69	5.81	5.86	5.90
	700	1.47	1.86	2.19	2.52	2.84	2.91	3.05	3.07	3.11
	800	1.64	1.93	2.14	2.48	2.76	2.93	3.52	3.69	3.71
Ni	900	1.75	2.25	2.61	2.93	3.29	3.55	3.57	3.55	3.52
	1000	1.82	2.57	3.71	3.52	4.07	4.48	4.57	4.61	4.62
	1100	2.34	2.79	3.39	4.00	4.55	5.00	5.34	5.48	5.48

Formulas for calculating the sputtering yield of monocrystals have been obtained theoretically by Martynenko et al. [62]. The dynamics of atomic collision cascade development in monocrystals leading to sputtering have been modeled on a computer by Shul'ga et al. [144].

Various experiments show that polishing metals introduces errors in determining the sputtering yield of less than 10% when the ion energy is about 500 eV [62]. At the same time, Rosenberg and Wehner showed that the sputtering yield of a nickel rod with a 0.45-mm pitch screw thread for 100-eV mercury ions is 0.4 times the atomic yield from a smooth nickel rod. At ion energies of 400-600 eV, however, the yields differ by 10%.

Studies of sputtering of W and Mo samples with different porosities (16-75%) by 200- and 500-eV Cd^+ ions showed that compared to rolled samples, the pores reduce the sputtering yield to 0.55 times the maximum value for a porosity of 40 to 75% [97]. In this case the metal grains differed greatly in shape and size. Another feature of the dependence of the sputtering rate of lanthanum hexaboride on porosity has been observed [98] during bombardment by 3- to 7-keV argon atoms and 7-keV ions from air. Under these conditions the sputtering rate rose or was practically constant when the porosity was varied over 5-30%. Further increases in the porosity led to a sharp, exponential growth in the sputtering rate. In the first case the reduction in the sputtering yield is explained by reverse deposition of sputtered atoms on the walls of pores. In the second, the sputtering behavior is explained by the onset of catastrophic destruction for porosities greater than 30% caused by reduced integral durability of the sample, which leads to enhanced removal of material. In a third case Borisenko, Dorofeev, and Pachinin [144] have observed increased sputtering of porous silicon with a volume density of 0.8-2.0 g/cm^3 compared to monocrystalline silicon during bombardment by 150-500-eV Ar$^+$ ions. This effect is explained by the reduced binding energy of atoms in porous silicon.

4.10. Sputtering Yield Dependence on Target Surface Temperature

The effect of the surface temperature of a sample on its sputtering yield is the subject of a rather large number of papers [62]. Arifov and co-workers have shown that when a surface is free of contaminants, the sputtering yields of tungsten and tantalum are independent of temperature up to 2000 K. The ratios of the maximum temperature to the melting temperature of these materials are 0.55 and 0.61, respectively.

Lebedev, Stavisskii, and Shut'ko [62] have made systematic studies of the dependence of the sputtering yield of a variety of metals (Fe, Ni, Ti, Pt, Nb, W, Ta, Mo, Re) and graphite on the energy of Cs$^+$ ions and the target temperature (Table 4.16). It is clear from Table 4.16 that the sputtering yield of iron increases by 30-40% and that of nickel, by 50-78% when the temperature is raised by a factor of 1.6. In this case the ratios of the maximum target temperature to the melting temperature are 0.76 and 0.80, respectively.

A stronger, exponential dependence was observed during sputtering of Bi, Zn, Ag, Cu, and Au by 45-keV xenon ions at a dose of 2.9·10^{16} cm^{-2} (Fig. 4.8), as well as during sputtering of electrolytic, rolled, and monocrystalline copper with a (101) plane by 400-eV Ar$^+$ ions [102]. Over the temperature range 600-1100°C the sputtering yield of electrolytic copper increases from 2 to 5. For the monocrystal, the temperature dependence is weaker, but the rise in the sputtering yield starts at temperatures above 850°C (from 0.5 at 700°C to 1.0 at 1000°C).

In extensive experiments on sputtering by various ions of silver, copper, and tantalum, Almén and Bruce [62] studied the temperature dependence of the sputtering yield of nickel, platinum, and silver over the range 20-800°C. A small peak in the sputtering yield of nickel is observed at 200-300°C. For sputtering by 45-keV krypton ions the peak is 7.3, while at 800°C the yield has fallen to 5. The sputtering yield of platinum decreases monotonically from 11 at 20-200°C to 8 at 800°C. During sputtering of silver the sputtering yield remains constant at 22 over the range 20-600°C and then begins a rapid rise to 25 at 820°C.

A sharp rise in the sputtering yield of ferromagnetic materials (nickel and gadolinium) at their transition from the ferromagnetic to the paramagnetic state has been reported [99]. In a narrow temperature region near the Curie point (about 30°C) the sputtering yield of monocrystalline gadolinium was a factor of 2 greater than in the ferro- and paramagnetic states. This effect is explained by a change in the binding energy of the atoms and in the propagation conditions for collision cascades.

Theoretical calculations [100] show that for a target temperature of roughly 0.7 times the melting temperature, the sputtering yield begins to rise exponentially. The same conclusion was reached earlier [101] from observations of a sharp increase in the sputtering yield of gold and silver at temperatures above 1000 and 900 K, respectively, or about $0.73 T_{melt}$ (Fig. 4.8).

The following formula has been proposed for the sputtering yield at low ion energies (below 100 eV) but with high ion current densities and high target temperatures on the basis of a model of the "hot spot" which appears at the site where an ion collides with an atom of the material [102, 144]:

$$Y = Y_0 \left[T + \kappa(E, T) \right] \frac{E^{-1/2}}{C^*} \exp\left[- \frac{U_0}{T + \kappa(E, T)E/C^*} \right], \tag{4.48}$$

where Y_0 and C^* are constants determined from the experimental dependence $Y(E)$ and κ is the energy accomodation coefficient of the ions.

Dolgov and Oranskii [122] have established that hollow cathodes of lanthanum hexaboride are rapidly destroyed at ion energies below the sputtering threshold. This effect is explained by the joint action of two factors: a high working temperature and ion bombardment. This complex mechanism is viewed as "temperature-stimulated cathode sputtering."

Zhdanov and Pletnev [144] have noted that at the melting temperature of a substance the enhancement in the atom energy is negligible compared to the binding energy in the lattice. But the process of getting the target atoms into motion is affected by the nature of the bonding of atoms in the lattice points, which depends on the temperature of the material. Increasing the target temperature leads to an increased contribution to sputtering from the first collisions in the atomic collision cascades. They increase the anisotropic escape angle distribution of the atoms. Calculations of the temperature dependence of the sputtering yield of iron and gold by D^+ ions show that for an ion energy of 0.4 keV there is a maximum of $Y(T)/Y(0)$ at temperatures $T/T_{melt} = 0$-0.2. For energies of 4 and 40 keV there is a negligible reduction in $Y(T)$ with rising temperature.

Kostyuk, Romanenko, and Bobryshev have shown [145] that at ion current densities of 10^6-10^{12} A/m^2 there is an increase in the sputtering yield (erosion) of molybdenum as the current density of H^+, Ar^+, and Xe^+ ions is raised. The duration of the heat flux and the type of ion have an effect on the erosion coefficient. For H^+ the sputtering yield is constant at about $2 \cdot 10^{-3}$ over 10^2-10^8 A/m^2 and then increases to 0.5 for 10-eV ions when the current density reaches 10^{12} A/m^2.

TABLE 4.17. Influence of Sputtering Parameters and Conditions on the Sputtering Yield

Parameter of the process	Range of variation		Y_{max}/Y_{min}
	parameter	Y (at/ion)	
Ion energy (eV)	$5\text{-}10^7$	$10^{-5}\text{-}10^3$	10^8
Charge on ion nucleus (rel. units)	1-83	$10^{-2}\text{-}55$	$5\cdot10^3$
Ion mass (amu)	1-209	$10^{-2}\text{-}55$	$5\cdot10^3$
Angle of incidence (deg): light ions:			
2 keV H$^+$ → Mo	0-85	$(2.2\text{-}22.5)\cdot10^{-3}$	10
8 keV H$^+$ → Mo	0-85	$(1.7\text{-}35)\cdot10^{-3}$	20
100 keV He$^+$ → Ni	0-80	$(4.58\text{-}29.6)\cdot10^{-2}$	6.5
medium and heavy ions:			
12 keV Ar$^+$ → Mo	0-74	3.2-20.8	6.5
8 keV H$^+$ → Cs$^+$	0-66	5.8-23.6	4.1
Ion momentum (g·cm/s)	$(8.5\text{-}36)\cdot10^{-12}$	0.2-12	60
Ion velocity (km/s)	20-60	4-12; 5-35	3-7
Nuclear charge of target atoms (rel. units)	3-83	1.1-22-64	20-58
Mass of target atom (amu)	7-209	1.1-22-64	20-58
Binding energy of target atoms (heat of sublimation, eV)	0.9-8.7	1.1-22-64	20-58
Gas pressure near target times target-substrate (vessel wall) distance pd (Pa·cm)	$10^{-2}\text{-}10^2$	1-0.01	100
Ion dose (cm^{-2})	$10^{17}\text{-}10^{19}$	$(1\text{-}16)\cdot10^{-3}$	16
Angle of rotation of the cleavage plane of a monocrystal (deg)	0-75	3.1-32.6	11
Target temperature (°C) metal			
8 keV Cs$^+$ → Ni	700-1100	3-5.3	1.7
45 keV Ar$^+$, Xe$^+$ → Ni	820-980	$Y_T=1\text{-}150$	150
pyrographite, 10 keV H$^+$ carbon composition,	20-800	$(6\text{-}200)\cdot10^{-3}$	35
10 keV H$^+$	20-800	$(5\text{-}20)\cdot10^{-3}$	4

oxide semiconductor [102a]			
100 eV $Ar^+ \to ZnO$ ionic crystal [102b]	20-230	0.06-0.41	7
300 eV $H^+ \to KCl$	20-400	5-50	10
Current density at target (A/m²) [102c]			
10 eV $H^+ \to Mo$ (theory)	10^8-10^{12}	$(2$-$500) \cdot 10^{-3}$	250
Cleavage plane of monocrystal:			
5 keV $Ar^+ \to Cu$ (111), (110)	$Y(111){:}Y(110)$	9.8:2.6	3.8
20 keV $O_2^+ \to Fe$ (111), (110)	$Y(111){:}Y(110)$	0.68:0.55	1.2
Atomic composition of molecular ions:			
70 keV Sb^+: 90 keV Sb_2^+:			
90 keV $Sb_3^+ \to Ag$	Sb_1^+: Sb_2^+: Sb_3^+	37:169:808	4-22
100 keV Bi^+:			
90 keV $Bi_2^+ \to Ag$	Bi_1^+:Bi_2^+	85:518	6
125 keV Bi^+:			
250 keV $Bi_2^+ \to Au$	Bi_1^+:Bi_2^+	$Y_{avg}=1900{:}3800$ $Y_{max}=13000{:}30000$	2 2.3
Surface roughness (mm):			
100 eV $Hg^+ \to Ni$	0-0.3	$Y_{smooth}{:}Y_{thread}$	2.5
400-600 eV $Hg^+ \to Ni$	0-0.3	$Y_{smooth}{:}Y_{thread}$	1.1
Surface porosity (%)			
200-500 eV $Cd^+ \to Mo$	10-75	$Y_{rolled}{:}Y_{porous}$	1.8
Ion current density for cleaning target of films (μA/cm²)	1-15-200	2-8-9.5	4-4.7
Magnetic phase transition at temperature (°C):			
15 keV $Ar^+ \to Ni$ (110)	350-370	2.2-3.3	1.6
Polymorphic transition at temperature (°C):			
10 keV $Ar^+ \to$			
(0001) α-Co → (111) β-Co [102d]	350-450	2-3.4	1.7

4.11. The Parameters of the Sputtering Process
and Their Effect on the Sputtering Yield

Thousands of studies have been devoted to research on cathode sputtering, and many thousands of papers deal with the technical and engineering applications of this physical phenomenon [63]. Since more than 20 parameters and conditions on which the sputtering yield depends are now known, it is necessary to evaluate, even if crudely, their relative importance (Table 4.17).

In concluding this chapter, we emphasize again that many complicated and varied phenomena occur during the interaction of ions with solids. Some of the interactions of ions with solids have been reviewed by Martynenko [103].

Chapter 5

ION CLEANING AND MILLING
OF SOLID SURFACES

5.1 Advantages and Features of Ion-Plasma Technology

The scientific, technical, and engineering foundations of the new, advancing vacuum ion-beam and vacuum ion-plasma technologies have been created by the work of many scientists and engineers during more than a century of comprehensive, in-depth research on the physical processes involved in the interaction of ions with solids and with the unremitting and ingenious development of ion sources and plasma heating devices for producing intense beams and jets of ions of various chemical elements. There are several fundamental distinctions which give certain advantages to ion-plasma processing.

The first fundamental advantage of vacuum ion-beam and ion-plasma technology over traditional methods of machining by cutting is the removal of material by fluxes of fast ions and atoms rather than by cutting tools, drills, milling cutters, and grinding wheels. Ion fluxes sputter materials of arbitrary hardness and strength, mainly in the form of atoms. This makes it possible to obtain very clean surfaces and all sorts of thin films made of metals, alloys, dielectrics, semiconductors, and new materials, such as superconductors with record high superconducting transition temperatures. Coatings obtained by cathode sputtering have great strength, adhesion to the substrate, and uniformity of composition and properties [62].

Cathode sputtering can be used to reveal the structure of all kinds of materials by ion etching.

The production of clean surfaces, thin film components of microelectronic circuits, and ion etching (engraving and milling) of microstructures have found extensive application in microelectronics and the production of semiconductor devices [104-108]. The extensive utilization of the technologies of ion cleaning of surfaces, ion etching of channels in complicated geometric shapes, and ion sputter deposition of wear-resistant and lubricating coatings make it possible to manufacture effective and cheap precision parts for gas bearings with the required surface and volume properties [109].

A second advantage of ion-plasma technology, which distinguishes it from the traditional technology for alloying steels and doping alloys and semiconductors by means of diffusion, is the rapid, often instantaneous, implantation of ions of arbitrary elements in any solid to obtain concentrations of the dopant elements in excess of their equilibrium solubility limit. Ion doping of semiconductors has found wide industrial application [110-113].

The nitriding of parts of machine tools and cutting instruments has been introduced in eight machine tool manufacturing plants in the USSR [114]. The advantages of this process over classical gas nitriding [114, 115] are the following:

1) a shortening by a factor of 2-4 of the entire cycle for nitriding pieces to obtain layers with effective thicknesses of up to 0.3-0.4 mm by reducing the times required to heat and cool the pieces and the isothermal exposure time;

2) the possibility of saturation at lower temperatures (beginning at 350-400°C), which ensures less deformation of the parts; then the surface roughness is kept within the limits $R_a = 0.63$-1.25 μm, so that in most cases nitriding can be performed as a final processing operation;

3) an enhancement in the plasticity of the nitrided layer and in the impact strength of the pieces;

4) extensive possibilities for controlling the saturation process, making it feasible to obtain nitrided layers with different structures and phase compositions and, therefore, to obtain pieces with specified mechanical and operational properties;

5) simple and reliable protection of surfaces that are not subject to nitriding;

6) reductions in the specific expenditure of electrical energy by factors of 1.5-3 and of gas, by factors of 20-50; and

7) complete ecological safety.

The shortcomings of the ion nitriding process under glow discharge conditions include:

1) the difficulty of nitriding deep holes;

2) the difficulty of placing pieces with highly different configurations and masses in the same enclosure;

3) the relative complexity of the equipment; and

4) the need for strict maintenance of the process operating parameters.

Ion nitriding of machine tools made of high-speed steels makes it possible to obtain nonbrittle diffusion layers of thickness 8-20 μm with a hardness HB = (7.4-7.8)·10^7 N/m^2, which increases their durability by a factor of 1.5-2.

A third fundamental advantage of vacuum ion-beam technology and the difference between it and Ohmic heating of plasmas in magnetic confinement devices lies in the more efficient transfer of energy from megawatt beams of fast deuterium ions to the ions in high-temperature deuterium–tritium plasmas at possible total injection powers of many hundreds of megawatts (see Chapter 3).

A fourth fundamental advantage of vacuum ion-beam technology compared to traditional engineering processes is the extremely high energies (up to 7000 MeV/nucleon) of heavy multiply charged ions [116]. Ion beams of this type can be used to dope thick material samples and change their electrical and physical properties. One extremely promising application for these beams is the manufacture of "nuclear" filters with pore sizes ranging from 3 nm to several tens of microns. These filters can be used to remove bacteria from water, milk, beer, and juices, to separate protein from serum, etc. Nuclear filters with hole diameters smaller than half the wavelength of incident radiation can theoretically completely (and in practice to a great extent) reflect the radiation.

Ion bombardment of glasses changes their refractive index. By irradiating thin optically transparent films with ion beams in accordance with a special program, it is possible to change their refractive index to make light guides of different thicknesses and configurations. Film light guides of this type are used in many branches of laser technology and in the manufacture of optoelectronics.

Introducing additives of elements with magnetic properties to films makes it possible to create components with large volume-distributed memory, which greatly extends the prospects for computer technology. Thin magnetic films that have been processed with ion beams improve the characteristics of memory devices and logical elements based on the retention and transfer of magnetic bubbles.

Irradiation with heavy ions may increase (or lower) the critical temperature for superconductivity. Beams of high-energy heavy ions are an important method for simulating radiation damage caused by fast neutrons. Ion radiography can be used to record density changes of less than 0.03 g/cm^2 with a spatial resolution of 1 cm. The intensity of the x-rays produced by heavy ions is roughly a million times greater than that induced by beams of electrons or protons. In addition, during irradiation by heavy ions there is practically no bremsstrahlung background. These factors significantly enhance the accuracy and sensitivity of the measurements.

The biological activity of heavy ions results in practically 100% deactivation of bacteria with specific energy losses of 10 MeV·cm^2/mg, i.e., with ion energies above 0.1 MeV/nucleon. Highly energetic ions have the advantage of transferring the bulk of their energy to a very small (diseased) part of the body at the end of their path, while x-rays, γ-rays, electrons, and neutrons also subject healthy cells in the human body to substantial and extremely harmful irradiation.

TABLE 5.1. Engineering Processes and Areas of Present and Future Application of Ion Flows and Beams [125]

Engineering process	Areas of present and future applications
Decontamination of surfaces	Electrovacuum industry Electronics and microelectronics Optical industry Production of rolled stock Machine building Space technology Physical chemistry and engineering research
Production of thin films of various materials	Semiconductor industry Electronics and microelectronics Electrovacuum industry Mass production of components Production of rolled stock Food industry Machine building Medicine

Doping and amorphization of
  semiconductors, metals, and
  dielectrics to improve the
  physical, mechanical, and
  chemical properties of their
  surfaces, hardness, durability,
  corrosion resistance, etc.
	Radiation technology
	Optical industry
	Metallurgical industry
	Medical and biological research
	Machine construction: automobiles,
	  tractors, agricultural machinery, etc.

Milling, drilling, and cutting
  of hard metals, alloys,
  insulators, and semiconductors
	Electronics and microelectronics
	Machine building (gas bearings, etc.)
	Optical industry

Grinding and polishing of
  glasses, metals, diamonds,
  and other solids
	Optical industry
	Machine building
	Production of precious stones

Elemental and isotopic analysis of solid surfaces	Semiconductor industry Electronics and microelectronics Instrument making Research in physical chemistry Hybrid and thermonuclear reactors
Production of intense fluxes of thermal and fast neutrons	Research in nuclear physics Radiation physics Radiation materials science

Simulation of radiation damage
  of structural materials in
  breeder, hybrid, and thermo-
  nuclear reactors
	Physics of radiation damage
	Radiation materials science

Production of new materials
  (superconductors, etc.) and
  metastable solid solutions
	Cryogenic technology
	Metallurgy

Production of atomic vapors for laser isotope separation	Metallurgy
Ion-beam therapy	Medicine

Heating of plasmas to
  thermonuclear temperatures
	Thermonuclear experiments and
	  future reactors

Simulation of sputtering and
  radiation erosion of surfaces
	Thermonuclear experiments and
	  future magnetically confined reactors

Welding of metals, alloys, and dielectrics	Machine building Electrovacuum and electronics industry
Pumping of gaseous-discharge lasers	Quantum electronics
Creation of a hydrogen maser	Space communications Medical and biological research

TABLE 5.2. Glow Discharge Operating Conditions for Cleaning Surfaces Prior to Film Deposition

Voltage	0.7-30 kV
Discharge current	15-600 mA
Current density	0.1-10 mA/cm^2
Gas pressure (argon, neon, helium)	53-1.3·10^{-2} Pa
Duration of processing	5-90 min

A fifth fundamental advantage of vacuum ion-beam and vacuum ion-plasma technology is the possibility of controlling the effects of low- and high-energy ions on the chemical and structural transformations at the surface and in near-surface layers. Radiation-induced chemical and thermal processing makes it possible to efficiently carry out the following processes: nitriding, siliconizing, cementation, production of diamond and diamondlike films, texturing of refractory metal coatings, and production of layers on metals and steels with enhanced hardness, durability, and corrosion resistance [117-137].

A sixth advantage of ion-beam technology is the possibility of directly monitoring the state and dynamics of changes in the surface as it is worked [138-140].

Finally, a seventh advantage of vacuum ion-beam and ion-plasma technologies is the possibility of automating and optimizing engineering processes [104], unlike, for example, in the case of chemical processing.

An important deficiency of ion-plasma technology is the need to set up and carry out engineering procedures in a vacuum. Vacuum engineering and technology, however, are already rather well developed, and in the future unrestricted vacuum conditions will be guaranteed in outer space, where space technology and production will develop extensively [127].

The problems of ion-plasma technology are discussed at various scientific conferences and have been summarized in monographs and reviews [122, 123, 133, 134, 141-174]. Table 5.1 lists the engineering procedures and possible areas for the use of high-current and dense ion and atom flows.

5.2. Ion Decontamination of Surfaces

The physical, electrical, and chemical properties of the surfaces of metals, semiconductors, and other materials (electron work functions, ion and electron emission, adhesion of films to substrates, corrosion resistance, etc.) depend to a great extent on their contamination by extraneous substances.

Of the many methods for decontaminating surfaces, the best results are obtained from decontamination by ion sputtering. Unlike most other methods, which act selectively on contaminants (greases, adsorbed water, gases, oxides, carbides, nitrides), ion bombardment is extremely effective at removing all of these and other contaminants. Thus, ion bombardment is used to remove contaminants from the surfaces of metals, alloys, semiconductors, and dielectrics of arbitrary hardness and adhesivity. Another advantage of sputter cleaning is that the process itself does not contaminate the surface and under certain conditions (low ion energies) does not cause significant damage to the layers of material near the surface.

Surfaces are cleaned by ion beams or by glow discharge plasmas (Table 5.2) [62].

The specifications for an ion flux used to decontaminate surfaces, including removal of films of residual gases in the vacuum system, have been described in Section 4.8. As an example, suppose it is necessary to remove a 1-μm-thick aluminum oxide film. The sputtering yield of Al_2O_3 bombarded by 10-keV Kr^+ ions [63] is 1.4 ± 0.2 atoms/ion. For an ion flux with a current density of 10 mA/cm^2, Eq. (4.1) gives a sputtering rate of $v_s = 10^{-5} \cdot 10^{-2} \cdot 1.4 \cdot 10^2 / 2.7 \approx 5.3 \cdot 10^{-6}$ cm/s. The time to remove an oxide layer with a thickness of 10^{-4} cm by ion bombardment is about 20 s.

Glow discharges are used to clean the surface of sheet steel during application of protective coatings. Plasma sources with closed electron drift and an extended accelerating region offer great promise for cleaning steel surfaces. The "wet" (chemical and electrochemical) methods traditionally used in metallurgy to clean rolled stock have several major deficiencies, including: incomplete cleaning and the presence of residual cleaning materials on the surface which are outgassing sources in a vacuum; a tendency for absorption of hydrogen at the surface by electrolytes containing constituents that are difficult to degrade biologically; and large amounts of production residues and expenditures of process water.

One such source, the UZDP-100, has been used for ion-plasma cleaning of the surface of samples of steel (08 KP) strip in a laboratory vacuum chamber. The rate and quality of vacuum ion-plasma processing were substantially (1.5-5 times) higher than with chemical cleaning. Thus, in the production cycle used at the Magnitogorsk Metallurgical Plant 1300-1350 mg/m^2 of contaminants are removed over 25 s by chemical cleaning or 1400-1500 mg/m^2 when an active substance, syntomide, is used. With ion-plasma processing this amount of contaminants is removed over 2-5 s [175]. After all processing is completed, the edge contact angle is 23-26° with an initial value of 52-55°.

These results indicate the appropriateness of testing ion-plasma cleaning at industrial metallurgical plants. Some metals form very durable films of oxides, carbides, and other compounds. For example, the heat of vaporization of boron in the atomic state is $5.9 \cdot 10^5$ J/mole, while the heat of dissociation of boron oxide is $2.6 \cdot 10^6$ J/mole. Heavy ion beams with energies of 20-50 keV are especially effective for removing such films.

Production of atomically pure surfaces. Farnsworth and Schlier have developed a technique for producing atomically pure surfaces [62]. It consists of repeated cycles of ion bombardment followed by annealing of the samples to remove impurities and implanted ions. The ion bombardment conditions are: argon pressure 0.13 Pa, accelerating voltage 200-600 V, current density 100 $\mu A/cm^2$, overall duration of irradiation 0.5-1 h. Titanium, nickel, silicon, and germanium are annealed at 500°C. With this method it is possible to reduce the level of contamination on the surface to roughly 5%. Electron diffraction studies showed that the structure was not damaged.

In one ion-optical system design for an ion beam (Fig. 5.1), the electrode system electrostatically focuses argon ion beams to diameters of 9.5 and 4 mm at energies of 100 and 600 eV and current densities of 8 and 240 $\mu A/cm^2$, respectively, at a sample located 60 mm from the edge of the last electrode.

Production of clean surfaces. An ion gun has been used in a slow-electron diffraction camera during studies of molybdenum, copper, and ceramic surfaces on substrates of yttrium oxide, thereby ensuring the production of pure surfaces under different operating conditions. Ion decontamination of the surfaces is used in all devices for producing films with the aid of diode, triode, radio-frequency, magnetron, ion-beam, and other systems. The ion sources developed for plasma heating in thermonuclear experiments can be used with argon, krypton, and xenon ions to clean large surface areas (1000-1500 cm^2). Cleaning a steel surface with a plane magnetron discharge prior to depositing nickel films of up to 5 μm in thickness has been found to be more effective and economical than glow discharge cleaning [177] (Table 5.3). In order to obtain a nickel film on iron with good adhesion, the magnetron discharge requires only 15% of the energy expenditure of a glow discharge.

Ion-plasma surface cleaning is widely used in microelectronics and in the semiconductor and optical industries.

To a major extent the creation of a clean surface for the first wall surrounding the high-temperature plasma in a magnetic confinement thermonuclear experiment determines the fraction of power lost by the plasma through impurity radiation. This power is 50-70% of the power applied to the discharge. In the T-10, T-11, PLT, ASDEX, and D-III tokamaks, a glow or low-power Taylor [161] discharge is used to clean the vacuum vessel surface. The energy flux to the surface is roughly 0.05 W/cm^2, the electron temperature of the plasma is 2-5 eV, and the electron density at the plasma center is 10^{11} cm^{-3}. The main constituents of the residual gas are water vapor at $(1-3)\cdot10^{-4}$ Pa and methane at $(1-13)\cdot10^{-5}$ Pa. At higher discharge powers (0.6 W/cm^2 and $T_e \sim 100$ eV) the residual gas is made up of methane and CO. In the first case $Z_{eff} = 2$-3 and in the second, 4-10. In doublet D-III the 70-m^2 surface is initially cleaned with a hydrogen discharge lasting roughly 500 h, and then for 25 h with an oxygen discharge [178].

Rutherford backscattering of 2-MeV $^4He^+$ ions can be used to measure light atom contamination on silicon surfaces at levels of 10^{15}-10^{16} $atoms/cm^2$ and heavy atom contamination of up to 10^{12} $atoms/cm^2$. If $^{14}N^+$ ions of the same energy are used, it is possible to measure 10^{-5} monolayer of heavy impurities [179]. This method was used to measure impurities near the vacuum wall in the Hungarian tokamak MT-1, where the size of the plasma column is controlled by a molybdenum limiter.

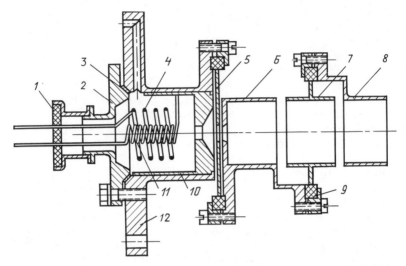

Fig. 5.1. The design of the ion-optical system for an ion gun used to obtain atomically pure solid surfaces [176]: (1) metal ceramic current feedthrough; (2) flange; (3) copper compression seal; (4) accelerating grid made of 0.5-mm-diameter tungsten wire; (5) extraction and (6)-(8) focusing electrodes; (9) precision ceramic ring; (10) discharge vessel; (11) linear channel cathode made of a 170-mm length of tungsten wire (ϕ = 0.2 mm) wound into a 5-mm-diameter spiral with a 1-mm pitch; (12) discharge vessel case.

TABLE 5.3. Comparison of the Parameters and Effectiveness of Glow and Magnetron Discharges for Cleaning Steel

Parameter	Glow discharge	Plane magnetron discharge
Voltage range, kV	1.5-4	0.3-0.6
Ion current density, mA/cm^2	0.5	37
Surface energy flux, W/cm^2	1.4	18.5
Pressure range, Pa	4-11	0.3-1
Rate of sputtering of steel, nm·cm^2/(W·s)	0.3	1
Thickness of film with good adhesion, nm	13	6
Energy release per cm^2 of substrate, J	43	6
Minimum substrate temperature for good adhesion, °C	100-120	60-80
Maximum possible cleaning rate (10-cm-long zone), cm/s	0.3	30

A plasma with an initial hydrogen pressure of $3\text{-}4 \cdot 10^{-2}$ Pa, a peak ion temperature of 100 eV, and an electron temperature of 500 eV was separated by a distance of 10 mm from the first wall. The contamination of 10×15 mm^2 monocrystalline silicon probes was measured within limits of ±1 mm. Probe A, located in the limiter shadow (d = 1 mm), picked up contamination from 203 discharges and probe B, which protruded 1 mm from the limiter, from 751 discharges. The state of the probe surfaces was studied after they had been exposed on both the electron and ion sides of the discharge. On the ion side the silicon became completely amorphous to a depth of 2-6 nm, and on the electron side thick layers with a low degree of amorphization of silicon monocrystals were observed. At the edge of the first wall the following surface densities were observed on the ion side of probe A: Fe 7-$9 \cdot 10^{14}$; Cr 2-$4 \cdot 10^{14}$; Mo $3 \cdot 10^{13}$; Cu $2.5 \cdot 10^{13}$; Si $2.5 \cdot 10^{16}$; C and O $8 \cdot 10^{15}$; P $1 \cdot 10^{15}$; Ca $5 \cdot 10^{13}$; and Cd $5 \cdot 10^{12}$ atoms/cm^2. On the electron side of the slabs the same elements were detected in roughly the same amounts: Fe $8 \cdot 10^{14}$ Cr $1.5 \cdot 10^{14}$; Mo $4 \cdot 10^{13}$; Cu 4-$6 \cdot 10^{13}$; Si $9 \cdot 10^{16}$; O $8 \cdot 10^{15}$; C $6 \cdot 10^{15}$; P $2.5 \cdot 10^{15}$; Ca $8 \cdot 10^{13}$; and Cd $2.5 \cdot 10^{12}$ atoms/cm^2.

About 80 methods and types of apparatus have been developed for analyzing the chemical composition, atomic and electronic structure, location of atoms in the crystal lattice, and relaxation of surfaces. Many of them are very complicated and expensive. At the same time, they do not allow analysis of large surface areas. Thus, a simpler method is to be recommended.

Penning, Moubis, and Jurrinas found in 1946 that under stationary glow discharge conditions, the cathode voltage fall of a particular material has a characteristic value after operation for an extended period [180]. For a constant pressure and discharge current, the voltage between the cathode and anode increases with the passage of time, reaching an asymptotic stationary value. Cathode sputtering removes an oxide layer and reduces the ion–electron emission coefficient.

In an anomalous glow discharge at constant voltage and pressure the discharge current decreases and reaches a stationary value. It has been found that for extremely pure argon (99.999%) at a pressure of 6.6 Pa and a constant voltage of 2 kV, the discharge current initially rises to about 75 mA and then falls to a roughly stationary value of 6.7 mA after 5 min. A simple soft x-ray spectrometer mounted in the discharge vessel showed that the principal impurities on an erbium cathode surface were carbon and oxygen. When the amount of oxygen on the cathode fell to a level of 0.01 after 50 min, the current had reached its stable value. This integral characteristic of the discharge current was used in ion deposition of thin aluminum films on a uranium surface [180].

A 3-kV, 90-mA anomalous glow discharge in oxygen running for 12-19 min completely sputtered a 100-nm-thick film of fatty acids with long chains of CH_3, $(CH_2)_n$, and COOH molecules with $M = 228$-397 amu deposited on 15.5-cm-diameter aluminum electrodes [181]. Argon at a pressure of 13 Pa only removed 60-80 nm of a 100-nm film through physical sputtering over this time. For an argon pressure of 133 Pa, at first the film was observed to thicken as a result of polymerization, but it subsequently underwent sputtering over this time to the extent noted. Hence, chemical sputtering was also involved.

Formation of polymer films during ion beam bombardment. Polymer films develop during bombardment of copper and stainless steel by H^+ and He^+ ions at a current density of 0.1 mA/cm^2 and energies of 50 and 500 keV, respectively, when the residual gas pressure is 1.3-$6.7 \cdot 10^{-3}$ Pa. The formula

$$n = n_0 + \beta t \left[1 - \exp(-\alpha I) \right] \tag{5.1}$$

has been proposed for describing the growth of surface contamination owing to proton bombardment [182], where n is the surface density of contaminant, n_0 is the initial surface impurity density prior to ion bombardment, β and α are proportionality constants, t is the duration of bombardment (s), and I is the current. An amorphous carbon-containing layer of thickness $2 \cdot 10^{17}$ atoms/cm^2 will reduce the sputtering yield of a copper surface for 500-keV He$^+$ ions by a factor of 100 from the value of $2 \cdot 10^{-2}$ atoms/ion for a pure surface.

Desorption (sputtering) of surface adsorbed layers by ion bombardment. Low-energy ion scattering [183] has been used to study the sputter desorption of H and CO from nickel and tungsten targets by 200-2000 eV He$^+$ and Ne$^+$ ions. Since the scattered ion intensity is proportional to the surface density of adsorbed atoms, the scattered ion current decreases exponentially with time as

$$I_A(t) \sim \exp(-i_0\, \sigma_D\, t), \qquad (5.2)$$

where i_0 is the ion-beam current density and σ_D is the desorption cross section. The desorption cross section can be estimated by measuring the current I_s^0 of ions scattered from a clean surface:

$$\ln(1 - I_s/I_s^0) = -i_0\, \sigma_D\, t + \text{const.} \qquad (5.3)$$

The cross section for desorption of CO from a (110) Ni surface by ^{4}He$^+$ ions rises from $4 \cdot 10^{-15}$ to $2 \cdot 10^{-14}$ cm^2 when the ion energy is raised from 400 to 1600 eV. There is a simple relationship between the sputter desorption cross section and the sputtering yield:

$$Y = \sigma_D\, n, \qquad (5.4)$$

where n is the surface density of adsorbate. For a single monolayer $n \approx 10^{15}$ cm^{-2}.

The desorption sputtering yield of a hydrogen monolayer on a (100) W layer for 500-1500 eV ^{3}He$^+$ ions has been calculated using the MORLAY program (Table 5.4). In Table 5.4 E_d is the minimum energy required to displace a target atom and E_c is the cutoff energy below which particles do not move in a collision cascade.

The sputtering desorption yield depends on the angle of incidence of the ions. For desorption of CO molecules from a (110) Ni surface by Ne$^+$ ions the yield is greatest at an angle of incidence of 70° (relative to the normal), and for desorption of sulfur atoms from a (111) Ni surface, at 55°.

The effect of the target atom and bombarding ion masses and of the binding energy of chemisorbed nitrogen atoms on the sputtering yield of two-component Mo (100) and W (100) targets has been studied using ion bombardment desorption by Winters [164]. The initial nitrogen atom density was about $6 \cdot 10^{14}$ cm^{-2}. The energy of Ar$^+$ and Xe$^+$ ions was varied over the range 300-5000 eV. It appears that the mass of the matrix atoms has an important effect on the sputtering yield, particularly at low ion energies. The ratio of the transverse sputtering cross sec-

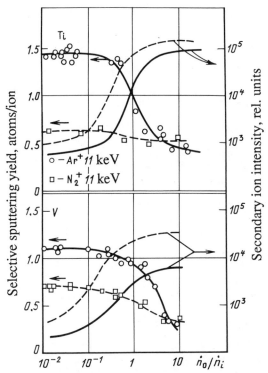

Fig. 5.2. Dependence of the sputtering yield and secondary ion emission intensity for titanium and vanadium in oxygen during bombardment by 11-keV Ar^+ and N_2^+ ions on the ratio of the fluxes of oxygen atoms and ions [185].

TABLE 5.4. Calculated Desorption Yields (atoms/ion) of Hydrogen Monolayers on (100) W Surfaces for a Binding Energy of 1 eV [183]

Ion energy, eV	Desorption yield, atoms/ion	
	$E_d = E_c = 5$ eV	$E_d = E_c = 1$ eV
500	0.20 ± 0.03	0.36 ± 0.04
1000	0.09 ± 0.02	0.25 ± 0.03
1500	0.09 ± 0.02	0.15 ± 0.02

tions $\sigma(N^+\text{-Mo})/\sigma(N^+\text{-W})$ varies from 0.46 at 500 eV to 0.65 at 5000 eV for xenon ions. The effect of the target mass is more important for heavier ions. For argon ions these ratios are 0.73 and 0.82, respectively. Neither chemical sputtering of adsorbed nitrogen atoms nor thermal sputtering made a perceptible contribution. The absolute cross sections for desorption of nitrogen by xenon ions at the maximum (3-4 keV) for Mo (100) and W (100), respectively, are $1.5 \cdot 10^{-15}$ and $2.3 \cdot 10^{-15}$ cm^2 or 0.9 and 1.4 atoms/ion.

TABLE 5.5. Effect of Cleaning Methods on the Coefficient of Friction [109]

Material of friction pair	Coefficient of friction after cleaning		
	washing in solvents	in solvent vapors	glow discharge
ShKh-15 Steel–ShKh-15 Steel	0.18	0.22	0.51
Al_2O_3 – Al_2O_3	0.23	0.28	0.87
ShKh-15 Steel–Al_2O_3	0.27	–	0.66

TABLE 5.6. Characteristics of the Microscopic Geometry of Aluminum Oxide Surfaces (with 95% confidence) before and after Sputtering [109]

Characteristic	Before processing	After processing
Arithmetic mean deviation, R_a (μm)	0.077 ± 0.2	0.626 ± 0.42
Greatest profile projection height, H_{pr} (μm)	0.012	1.1
Greatest profile height, H_{max} (μm)	0.24	3.8
Surface roughness, GOST 2789-73 standard	11	8
Etching rate (nm/min)	8.0 (initial)	9.3 (final)
Angle of inclination of roughness	$0° 15' \pm 0° 11'$	$10° 15' \pm 6° 54'$

500-eV Cs^+ ions appear to be an effective means of removing C and O from Ti, W, and Si surfaces and from thin (22 nm) films of Si_3N_4 on Si [184]. An approximately 1.2-nm-thick layer of these contaminants on a silicon nitride film was removed after bombardment by Cs^+ ions to a dose of 20 μA·min/cm². A 1-nm-thick layer of Si_3N_4 was sputtered by a current density of about 1 μA/cm² over 15 min. The sputtering yield was about 4 atoms/ion over the range 0.055-0.073 nm/(μA·min/cm²). These values were obtained for a sample temperature of 800°C and a pressure of 6.7·10⁻⁸ Pa. At room temperature the sputtering efficiency is maintained, but cesium ions remain on the surface.

The oxidation of titanium and vanadium during bombardment by 11-keV Ar^+ and N_2^+ ions has been studied by measuring the sputtering and secondary ion emission yields [185]. Ion beams with current densities of 100 μA/cm² and 99.999% pure oxygen were used with a residual oxygen pressure of 2.7·10⁻⁶ Pa (Fig. 5.2). The sputtering yield of these metals begins to decrease when the ratio $n_0/n_i \geq 0.1$. When this ratio is equal to 10, the sputtering yield of titanium bombarded by argon ions decreases to 1/3 of its initial value and that of vanadium is reduced by a factor of 4. For clean surfaces they are 1.42 and 1.7 atoms/ion, respectively. Under the assumption that TiO_2 and V_2O_5 are formed, the selective sputtering yields of titanium and vanadium are 4.2 and 3.9 atoms/ion, respectively. The increased sputtering yield is explained by a reduction in the binding energy of atoms of these metals in their oxides.

Ion-plasma surface cleaning is widely used in electronics, microelectronics, electrovacuum technology, optical fabrication, and in physical, chemical, and thermonuclear research.

TABLE 5.7. Roughness Class of Surfaces before and after Ion Etching [109]

Material	Before sputtering	After sputtering
ShKh-15 steel	14	9–10
Kh18N10T steel	12	8
Aluminum oxide	12	8–9

5.3. Ion Milling (Etching)

In machine building. A detailed study has been made by Grigorov and Semenov [109] of the possibility of using ion sputtering for the manufacture of gas bearings. Bearings with a gaseous lubricant were proposed and have found practical application in the USSR. They are used in a number of branches of industry. The low viscosity of gaseous lubricants makes it possible to employ them in equipment with shaft rotation speeds of up to 700,000 rpm. The reduced coefficient of friction in these bearings is on the order of 0.0001 or less. The specific load-carrying capacity of gasdynamic bearings may be as high as $0.98\text{-}1.96 \cdot 10^5$ N/m^2. Precision (fractions of a micron) fabrication of multiple-thread gasdynamic grooves with profiled indentations and high surface smoothness (9-10th class) in composite materials with wear-resistant and antifriction coatings is difficult with traditional machining techniques. Thus, a fabrication method that was new for machine building was proposed.

Three types of systems were used:

1) for ion sputtering of materials in a dc triode arrangement (with a heated tungsten cathode, an anode, and a sputtering target with substrates for deposition);

2) a "Plasmonic"-type double target for sputtering materials in an rf discharge; and

3) an ION-1V apparatus for dc and rf discharge sputtering that was modernized by adding a magnetic field for stabilized plasma formation and to ensure steady operation of the discharge.

Parts were initially cleaned in flowing hot soap and water, dried, and then cleaned successively in "Galosh" benzine solvent (GOST standard 443-76), carbon tetrachloride (or benzene, or ether), and, finally, alcohol. Final cleaning was in a vacuum glow discharge. Before this the working zone and component surfaces were cleaned with a vacuum cleaner to prevent entrapment of dust. The vacuum chamber was pumped to a pressure of 13-40 Pa. The discharge current was on the order of 10 mA at 600 V from a model UIP-1 standard dc power supply with a ballast resistance of 1280 Ω.

Table 5.5 shows the average coefficients of friction obtained after surfaces were cleaned by different methods. It is clear from this table that the most effective cleaning method is glow discharge sputtering.

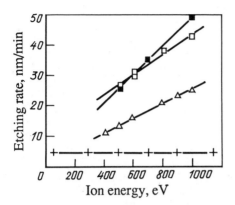

Fig. 5.3. The rate of chemical etching of Si in XeF_2 vapor (+), the rate of physical sputtering by Ar^+ ions (Δ) for a current density of 0.22 mA/cm², the rate of physical etching in the presence of XeF_2 vapor (□), and the rate of physical and chemical etching by Ar^+, Xe^+, XeF^+, and XeF_2^+ ions at a current density of 0.22 mA/cm² (■) [186].

TABLE 5.8. Comparison of the Technical Characteristics of Two Methods for Fabricating Gasdynamic Grooves on Ceramic Step Bearings [109]

Characteristic	Grinding	Ion etching
Shape of generatrix of grooves	Arc of circle, Archimedean spiral	Logarithmic spiral
Shape of end parts	Arc of circle	Theoretical
Deviation from generatrix (mm)	±0.2	±0.03
Deviation in depth (mm)	±0.003	±0.001
Surface roughness class	6	8
Number of grooves	≤9	unlimited

Ion etching (milling) of gas grooves was carried out on steel and ceramic step bearings which were attached to a multiple holder fixture. This fixture had a 1-mm-thick mask that covered the part of the surface of the piece (a plane or cylindrical shaft) that was not to be sputtered. All the pieces were made of Kh18N10T steel.

The optimum conditions for ion milling were the following: argon pressure about 1.3 Pa, accelerating voltage 2-3 kV, operating ion current density 0.5-1.8 mA/cm², discharge current 0.5-2 A, and magnetic field strength 5.57-7.16·10³ A/m.

The sputtering rates of stainless steel and aluminum oxide increased with the duration of ion bombardment. For the Kh18N10T stainless steel the sputtering rate can be estimated as (nm/min)

$$v_s = 90.0 + 0.0167 t \quad \text{for} \quad t = 0 - 600 \text{ min,} \tag{5.5}$$

and for aluminum oxide, as

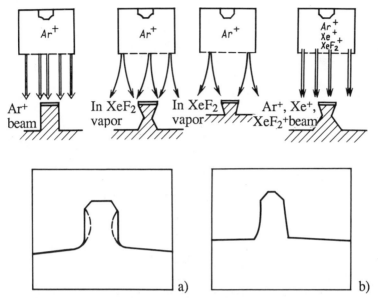

Fig. 5.4. Profiles obtained using a two-step process with an approximately parallel Ar^+ ion beam and a diverging Ar^+ beam in the presence of XeF_2 vapor (a) and profiles obtained using a two-step process with a diverging Ar^+ beam in the presence of XeF_2 vapor and roughly parallel beams of Ar^+, Xe^+, and XeF_2^+ ions (b) [186].

$$v_s = 8.1 + 0.0021\, t \quad \text{for} \quad t = 0 - 1000 \text{ min.} \tag{5.6}$$

During sputtering the microscopic geometry of the surface changes, most noticeably in the angle of inclination of the microscopic roughness profile (Table 5.6).

Fully satisfactory surface quality can be obtained by ion milling of gas grooves (Table 5.7). This ion milling technology makes it possible to fabricate gas grooves with optimum shapes, particularly, logarithmic spirals. The carrying capacity of gasdynamic bearings with grooves shaped in logarithmic and archimedean spirals is considerably (2-5 times) greater than for grooves with straight segments.

Ion milling is compared with traditional technology in Table 5.8. The ability to carry out successively the operations of ion etching of grooves, substrate and target cleaning, and deposition of wear-resistant and lubricating coatings (molybdenum disulfide and Ftoroplast-4) with simultaneous ion sputtering of tens and hundreds of parts at rates of up to 1 µm/min without breaking vacuum and shifting parts makes this process suitable for widespread use.

In microelectronics. Ion etching of microscopic structures has already found industrial application in microelectronics, where plasma-chemical and ion-chemical etching are used [104]. In 1965 submicron resolution (0.25 µm) was obtained on Au–Pt metal layers using electron lithography and ion etching [104].

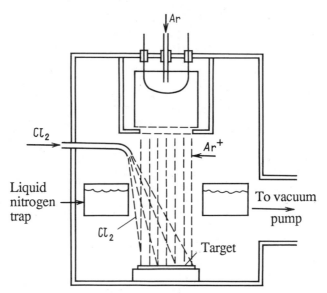

Fig. 5.5. Apparatus for chemical etching assisted by an ion beam [188].

In microelectronics the use of ion-chemical etching for removal of SiO_2 windows in type URMZ.279.045 devices [107] yields the following savings in chemical reagents per 1000 silicon wafers (kg): nitric acid 33.3, phosphoric acid 96, sulfuric acid 31, and glacial acetic acid 216. Besides the direct savings on chemical reagents, substantial savings are realized because of lower costs for labor and for the raw materials needed to neutralize the industrial effluents of strong acids and other materials, as well as saving large amounts of deionized water (tens of thousands of cubic meters per year for a single system).

Ion-chemical etching. Beams of ions of chemically active molecules, such as XeF^+, XeF_2^+, and CF_3^+, can be used to greatly augment the rate of physical sputtering by combining it with chemical sputtering. The effect is not simply the sum of the etching rates, but an intensification of the chemical interaction. The following forms of ion-chemical sputtering are, therefore, distinguished: reactive ion etching (RIE) in a plasma, reactive ion-beam etching (RIBE), and chemically assisted ion-beam etching (CAIBE) [186]. The relationship of the different types of chemical and physical etching is illustrated in Fig. 5.3 with silicon etching as an example. The etching rate of silicon in XeF_2 vapor at a pressure of about $1.3 \cdot 10^{-2}$ Pa was 4.8 nm/min. Bombardment by 1000-eV argon ions increased the sputtering rate by about a factor of 5. Bombardment by argon ions in the presence of XeF_2 vapor increased the sputtering rate by about a factor of 9 compared to purely chemical etching. Putting XeF_2 in the ion source led to an increase of a factor of 10 in the etching rate (up to 50 nm/min) compared to purely chemical etching. By using an ion source with a beam diameter of 15 cm and controlling the species composition, divergence angle, and ion energy, it is possible to obtain different profiles during etching of silicon covered with a photoresistive mask (Fig. 5.4).

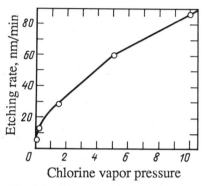

Fig. 5.6. Dependence on the Cl vapor pressure of the rate of chemical etching of GaAs assisted by 500 eV, 20 μA/cm^2 Ar$^+$ ion beam bombardment [188].

TABLE 5.9. Sputtering Rates of Optical Glasses

Optical material	Sputtering rate, μm/h		Optical material	Sputtering rate, μm/h	
	In argon	In CF$_4$		In argon	In CF$_4$
Quartz Glasses:	1.0	3.6	OFZ	1.2	3.6
K8	1.2	5.4	TF10	1.1	3.6
FK14	1.8	4.2	TF1	1.2	3.0
FK1	1.3	4.6	STKZ	1.0	3.0
OF1	1.2	3.6	LK5	1.2	4.0

Table 5.9 shows the sputtering rates of a number of optical materials in argon and freon rf discharges (frequency 440 kHz, ion current density 2-4 mA/cm^2, pressure 0.4 Pa) [187]. The rate of sputtering of quartz and KV glass decreases from 9 to 3.5 μm/h as their surface temperatures are raised from 50 to 200°C.

Extremely high chemical etching rates (5-10 μm/min) of GaAs in chlorine vapor at a pressure of 1.3 Pa have been obtained during bombardment by 1-2 keV argon ion beams with an ion current density of 1 mA/cm^2 [188] (Fig. 5.5). A liquid nitrogen trap located near the target produces a pressure drop of two orders of magnitude between the Kaufman type ion source and the target. The pressure of the reactive gas (chlorine) rises to 2.7 Pa near the target, but is kept at a level of $1.3 \cdot 10^{-2}$ Pa in the source region. At a pressure of 1.3 Pa the etching rate was 85 nm/min (Fig. 5.6). This means that one Ar$^+$ ion removed roughly 500 target atoms. When the ion energy flux at the surface was raised from 10 mW/cm^2 (20 μA/cm^2 at 500 eV) to 2 W/cm^2 (1-2 keV, 1 mA/cm^2), the rate of reactive etching reached 5-10 μm/min. This means that every fourth or fifth chlorine atom participated in a chemical reaction if we assume that the reaction products were GaCl$_3$ and AsCl$_3$. At such high etching rates, however, the etching was no longer anisotropic.

TABLE 5.10. Estimated Ranges of Ions and Energy Losses through Nuclear Collisions in a Silicon Target

Ion type and beam energy (keV)	Projected path R_p (nm)	Standard deviation in path ΔR_p (nm)	Transverse spread Δx (nm)	Energy loss (eV/nm)
Ga$^+$, 50	33.7	8.3	7.9	13.4
In$^+$, 30	19.0	3.3	3.9	18.0
In$^+$, 50	27.6	4.8	5.5	20.0
Sn$^+$, 50	27.5	4.6	5.4	20.3
Sn$^+$, 100	46.7	7.6	8.9	22.0

At lower ion energies and etching rates the surface damage was minimal and the aspect ratio (height to width of the step) exceeded 35. A grating with a spacing of 500 nm and a depth of 1.5 μm was obtained. A 2.5-cm-diameter Kaufman source with a graphite grid consisting of a set of cylindrical apertures was used. For a source–target distance of 19-44.5 cm and a beam divergence angle of 4-1.5° the etching uniformity was ±5% over a surface area of 4 cm^2 on a 5-cm-diameter wafer.

Production of gratings and structures with dimensions of 40-100 nm. Methods for producing gratings with line thicknesses of up to 40 nm were developed before 1983 in a number of American and Japanese laboratories [165]. These methods are based on ion and electron beams with diameters of 0.5-50 nm, lithography using these beams, and a method for doubling the number of lines.

In a scanning ion microscope at the University of Chicago, P.H. LaMarche et al. have obtained gallium ion beams with energies of 40-50 keV, diameters of 0.05-0.1 μm, and currents of 10-100 pA from a liquid metal ion source [165]. Electrostatic focusing was used to obtain these beams. A beam could scan a 0.4 × 0.4 mm^2 surface area. Silicon was used as a negative resistor with a sensitivity of 1 μC/cm^2. The fastest scan rate with a clearly distinct line width of 0.1 μm was 4 mm/s and corresponded to an ion dose of 1.5·10^{13} cm^{-2}. Optical gratings with up to 40,000 lines/cm were obtained at ion doses of 1.6·10^{14} cm^{-2} with subsequent etching in a solution of NaOH for 2-3 min.

In the Electrotechnical Laboratory at Ibaraki, Japan, liquid metal ion sources have been used to produce gallium, indium, and tin ion beams with energies of 30-50 keV, diameters at the crossover of 40-70 nm, and emission currents of 4-12 μA [166a]. The calculated range and energy losses of these ions are listed in Table 5.10. Etching of silicon begins at a critical dose of 3-5·10^{-6} C/cm^2. The implantation depth increases rapidly with dose, reaching 40 nm for ions with energies of 50 keV at a dose of 10 μC/cm^2. For a linear dose of 2.2·10^{-10} C/cm gratings with a width and depth of 50 nm and a spacing of 62-150 nm have been obtained with a 50-keV gallium ion beam. The etching depth is roughly equal to the sum $R_p + \Delta R_p$. The minimum channel width obtained was 26-30 nm.

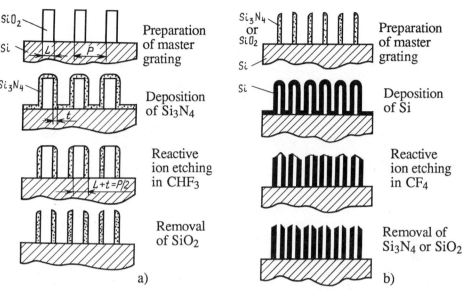

Fig. 5.7. The sequence for making twice the number of lines in an SiO_2 grating by making a grating of Si_3N_4 (a) and the same with an initial grating of Si_3N_4 or SiO_2 (b), with a final grating of Si [189].

At the MIT Lincoln Laboratories G.H. Randall et al. have used ion lithography to make template masks in nonstoichiometric silicon nitride with line widths of 80 nm and a grating period of 160 nm [166a]. A mask had a 1-μm-thick metal base on which a silicon nitride membrane with a thickness of 0.5-1 μm was deposited. The base and membrane were deposited on a 25-μm-thick silicon wafer. Using holographic indicial masks and contact x-ray lithography, a transparent pattern of 80-nm-thick nickel was deposited on the mask. Holes were etched in the nickel mask by reactive ion etching in CHF_3 at a bias potential of 500 eV. After a resistive mask of PMMA was applied, plasma etching was used to make windows in the silicon nitride membrane. The silicon wafer, whose cross section was covered by a 20-nm-thick nickel film, was etched from the opposite side. The PMMA was exposed to a 50- or 100-keV proton beam. This beam produces high contrast and requires less exposure time than an electron beam or x rays. The appearance (development) time is related to the absorbed energy by

$$R = CD^\alpha, \tag{5.7}$$

where R is the development rate (nm/s), C is an empirical constant, D is the absorbed dose (J/cm^3), and α is the contrast. It is found from a log–log plot of Eq. (5.7). For a proton beam in PMMA with a molecular mass of 950,000, $\alpha = 4$ while for 20-keV electrons $\alpha = 2.8$. Grating lines with a width and spacing of 80 nm were obtained in 300-nm-thick PMMA with a 0.5-μm-thick silicon nitride membrane irradiated by 100-keV H_2^+ ions to a dose of $2 \cdot 10^{13}$ cm^{-2}.

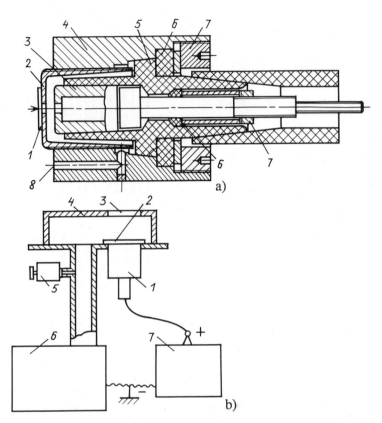

Fig. 5.8. A Rozenfeld–Makarov ion gun for drilling micron apertures by means of sputtering (a) [62]: (1) Piece to be drilled; (2) cathode; (3) anode; (4) case; (5) insulator; (6) rubber seal; (7) screw to compress seal; (8) channel linking discharge gap with vacuum system. A sketch of a system for ion drilling (b): (1) ion gun; (2) work piece; (3) observation window; (4) vessel cover; (5) leak valve; (6) vacuum pump; (7) high-voltage rectifier.

A method for sequential doubling of the number of lines in a grating to a period of 40 nm has been developed in a number of laboratories [189]. The essence of the method is shown in Fig. 5.7. Primary (master) gratings with a period of roughly 160 nm have been obtained at MIT Lincoln Laboratories by means of holographic lithography at 351.4 nm in a liquid medium. Coating deposition techniques were used in order to obtain x-ray masks with precision control of the width-to-period ratio of the lines. Then reactive ion etching in CHF_3 was used to obtain rectangular profile structures in SiO_2 with the exact sizes of the line widths. Then chemical vapor deposition was used to obtain an Si_3N_4 layer of the exact thickness. The upper surfaces of these layers were removed by reactive ion etching. Finally the SiO_2 between the Si_3N_4 windows was removed by isotropic selective etching. An x-ray shadow mask was made of 4.5-nm-thick carbon. Si_3N_4 gratings with a spacing of 40–240 nm were obtained on a surface area of about 1 cm^2. There is hope that gratings with a period of 10 nm can be obtained with the aid of primary gratings having a period of 40 nm.

TABLE 5.11. Speed of Drilling in Various Materials by Ion-Beam Sputtering

Material	Layer thickness, μm	V, kV	Discharge current, mA	Drilling time, s	Reaming time, s	Hole diameter, μm
Tantalum	75	15.5	0.44	440	10	14
Nichrome	30	15	0.10	100	30	28
Molybdenum	60	19.5	0.16	290	120	40
1Kh18N9T steel	200	13	0.18	52 min	30	7
Diamond	500	20	0.20	38 min	–	40

Equipment for ion milling. Many companies in various countries sell equipment for ion-beam milling [190]. We now examine some of this ion etching and milling equipment.

A *bench-top attachment for the* model IBT 200 *ion-beam thinning apparatus* of Edwards Ltd. for thinning solid materials with an ion beam at a speed of 1-15 μm/h down to a thickness of 50 nm or less has been installed in 306 coating systems from this company. It can be used to obtain very thin samples of metals, metallic alloys, ceramics, semiconductors, minerals, polymers, etc., in any laboratory that has a 305-mm-diameter deposition system [191].

An ion gun for making diaphragms with micron apertures was developed by L.B. Rozenfeld and A.I. Makarov in 1961, and in 1968 they developed the "Mikron" system for ion boring of holes with diameters from 2-3 μm to 300 μm in plane and cylindrical samples (Fig. 5.8). The gun operates as a gaseous-discharge canal-ray ion source. The anode is a cylindrical hollow of diameter 7 and length 22 mm covered by a diaphragm with a 3-mm-diameter hole. There is a hole in the cathode for ions to enter the vacuum. The pressure in the discharge gap is 13 Pa, which ensures autofocusing of the ions into a very thin beam. The beam is supplied by a very simple 12-25 kV rectifier with a current of up to 2 mA without any filters or voltage regulators. The discharge current is limited by a 100 kΩ resistor. The ion drilling speeds for different materials are listed in Table 5.11.

A *microion milling machine* with a line resolution of 50 nm has been developed by Technics (USA). It can be used for ion polishing and for deposition of metals, ceramics, or a combination of both at temperatures below 140°C.

Equipment for ion-beam milling with continuous mass analysis has been developed by the Commonwealth Scientific Corporation (USA). It is supplied with an automatic control for the sputtering rate with an error of less than 1 nm and shuts off automatically when the desired etching depth in the film has been reached. The milling speed is 1 μm/h or greater. Samples with diameters of 50 mm can be worked.

Kaufman source with electron bombardment. Of the ion sources employed for industrial purposes, this type has gained most widespread use. It was developed as an ion rocket motor in 1961 [171]. It uses a Penning discharge. Electrons are emitted by a cathode mounted on the axis of a cylindrical discharge chamber located in a weak longitudinal magnetic field (Fig. 5.9). Part of the cylinder serves as

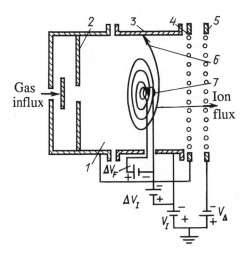

Gas
influx

Fig. 5.9. Diagram of a Kaufman ion source with a Penning discharge [171]: (1) ionization chamber; (2) screen; (3) anode; (4) extraction electrode; (5) accelerating electrode; (6) electron trajectory; (7) electron emitter.

an anode. The electrons move in the crossed magnetic and electric fields along spirals around the axis, so that their path length and the probability of ionization of the working gas are increased. The ions are extracted, accelerated, and focused by a two- or three-electrode ion-optical system which creates numerous elementary cylindrical beams. A 3-keV mercury ion beam with a current of 130 mA and average current density of 1.6 mA/cm^2 was obtained for a discharge current of 1.7 A, discharge voltage of 50 V, and magnetic field of $2 \cdot 10^{-3}$ T. The specific impulse was 5500 s and the overal energy efficiency 70%.

A magnetic field was introduced inside the chamber in order to further raise the source efficiency [171]. An argon ion current density of 1 mA/cm^2 was obtained with an accelerating voltage of 500 V. The 0.5-mm-thick extraction and accelerating grids had 2-mm-diameter holes that occupied 65% of their surfaces. The distance between the electrodes was 0.75 mm, and the pressure in the discharge vessel was 0.1 Pa.

Single-electrode ion-optical system. This system is a further improvement of the Kaufman source. The distance between the plasma and the electrode is 0.1 mm instead of the usual 1-2 mm. This makes it possible to increase the ion current density by more than a factor of 40 to as high as 20 mA/cm^2 at low energies. The use of permanent magnets which create a dipole magnetic field made it possible to develop a compact ion source with a beam diameter of 3 cm and an outside diameter of 6.4 cm [171]. An important advantage of this small source is the fact that it can be rapidly replaced with the use of a plug-in adapter. A compact ion source makes it possible to combine ion processing with other vacuum techniques, such as evaporation. One important disadvantage of these ion sources is the relatively short lifetime of heated cathodes (several tens of hours). Using a *hollow cathode* in the form of a tantalum pipe with an inner diameter of 12.7, wall thickness of 0.51, and length of 57 mm that has a pure tungsten diaphragm with a 1.5-mm-long, 1-mm-diameter aperture raises the operating lifetime of the cathode to 306 h in an ion source with a strong multipole magnet [165].

Chemically active gases rapidly destroy cathodes that are heated to high temperatures. In order to increase the continuous operating time and reduce metallic impurities in the ion beam, a 26-cm-diameter source has been developed in which the electron cyclotron resonance is used to create the discharge plasma [164]. Microwave power at a frequency of 2.45 GHz is delivered to the source through a ceramic window. Three magnetic windings create a field of $8.75 \cdot 10^{-2}$ T. A stable plasma is generated at a pressure of 0.0027-0.027 Pa in C_4F_8. The extraction and accelerating grids are made of 0.5-mm-thick molybdenum plates. 3900 holes with diameters of 3 mm were drilled in them. The distance between the grids is 2 mm. An accelerating potential of 500-1000 V is applied to the accelerating electrode, 200 V to the extraction grid, and the sample is kept at ground potential. Under these conditions a well-focused beam is obtained. The plasma density is $3 \cdot 10^{10}$ cm^{-3}, and the electron temperature is 3 eV. At an ion energy of 1 keV the current density reaches 300 μA/cm^2 with an inhomogeneity of $\pm 8\%$ within a diameter of 200 mm. The main components in the ion beam are CF_3^+, CF_2^+, CF^+, $C_2F_4^+$, and $C_3F_5^+$. The maximum etching rate for silicon dioxide is 70 nm/min, and for silicon, 30 times lower. The source can operate for about 40 h, after which a film produced by polymerization of C_4F_8 must be cleaned from the walls.

Ion and atom sources with currents of up to 10 mA and energies of up to 10 keV for narrow and wide beams for cleaning surfaces, etching, physical and reactive sputtering, and ion-stimulated sputtering of films are manufactured by Ion Tech. Ltd.

Accelerators, ion sources, negative ion injectors, neutron generators, and sputtering systems with ion energies from 200 eV to 10 MeV are built by the General Ionex Corp.

Dc diode sputtering systems, rf sputtering systems with a bias potential on the samples, and magnetron sputtering systems are often used in laboratory and industrial establishments.

The Austrian firm Xennon produces a series of fully automatic systems for dry etching of single wafers with diameters of from 50.8 to 127 mm with a structure resolution of better than 0.5 μm: XPL01P, XPL24, XPL26, XPL01RD, XPL6, and XPL SMED.

A fully automatic system for dry etching (model DEA-501) is manufactured by the Japanese company NEVA. Rf sputtering can be done with a resolution of 0.2 μm in the topology. The etching rate for molybdenum and silicon in CCl_2F_2 at 13 Pa and a power of 500 W can be as high as 400 nm/min with 0.7 W/cm^2. The error in etching is less than $\pm 10\%$ over wafers with diameters of 76.2 mm.

The model A400 VL laboratory system of Leybold-Heraeus can perform operations separately or in combination: ion plating, sputtering, and etching, deposition, and sputter deposition.

In most commercial ion etching systems, several semiconductor wafers are processed simultaneously. It is often difficult to obtain uniform etching of even a single piece, much less all of them, in the vessel. Thus, in some systems only one wafer is processed at a time. The productivity of these systems, however, is low. In order to combine the advantages of systems of this type, a new multichamber plasma etching system has been developed in Japan [192]. This system consists of six vessels, each of which is intended to carry out a certain operation. A wafer is sequentially moved through all the vessels, in which six wafers are being processed continually. The system can handle 40 wafers with diameters of 75, 100, or 125 mm in 1 h. The etching uniformity for single and multiple wafers is ±5%. The base pressure is $1 \cdot 10^{-4}$ Pa and the working pressure, 2-20 Pa. 120 W of rf power at a frequency of 13.56 MHz is applied to a single electrode. The etching rate is 100 nm/min for aluminum in SF_6 at 3 Pa and 60 W; 100 nm/min for polysilicon in $CF_4 + H_2$ at 6.65 Pa and 50 W; and 35 nm/min for silicon dioxide in $CF_4 + H_2$ at 6.65 Pa. The resolution for etching of fine pieces is 1 µm.

The experimental etching rate for silicon in a CF_4/O_2 plasma is in good agreement with the rate of chemical reactions between fluorine and silicon atoms [193]:

$$R_{F(Si)} = (2.91 \pm 0.2) \, 10^{-13} \, T^{1/2} \, n_F \, \exp(-1.73 \cdot 10^{-20}/kT), \qquad (5.8)$$

where $R_{F(Si)}$ is the etching rate (nm/min) for silicon, T is the temperature (K), n_F is the density of F atoms (cm^{-3}), and k is Boltzmann's constant ($1.38 \cdot 10^{-23}$ J/K). A study of the rate of etching of silicon dioxide in fluorine-containing gases such as NF_3 and ClF_3 in the presence of oxygen or fluorocarbons yielded a formula for the rate of etching of SiO_2:

$$R_{F(SiO_2)} = (6.14 \pm 0.5) \cdot 10^{-14} \, n_F \, T^{1/2} \, \exp(-2.61 \cdot 10^{-20}/kT). \qquad (5.9)$$

In conclusion, we note that a bibliography on ion milling (ion-beam etching) covering the period 1954-1975 has been published [194].

We now examine some possible industrial applications of ion milling.

Industrial applications for ion milling (machining of materials with ion beams). *Ion micromachining of memory devices based on cylindrical magnetic domains.* Ion processing has been used to remove surface and near-surface defects and contaminants resulting from abrasion working of orthoferrites. Two argon ion beams, located horizontally opposite one another, were directed onto an $Sm_{0.55}Tb_{0.45}FeO_3$ surface at an angle of roughly 75° to the normal. The sample was rotated about the normal at a rate of 15 rpm [195]. For an ion-beam current of 120 µA and accelerating voltage of 6 kV, the etching rate was 0.5 µm/h per side. The etching rate can be increased by a factor of 100. The layer to be removed was from 400 to 2500 nm thick. Roughly 90% of the surface was perfectly flat, ±300 nm (the limit of accuracy of the instrument), after ion bombardment. After processing, the samples are characterized by polished surfaces and the required magnetic domain patterns. Ion machining makes it easy to obtain parallel grooves, concentric circles, ridges, or valleys, and many combinations of these.

Use of ion beams for smoothing surfaces. For this purpose a surface containing prominences and depressions is coated with a layer of material that has the same sputtering yield as the work surface at large angles of incidence of the ions (60°-70° to the normal). The thickness of the coated layer must be considerably greater than the depth and height of the nonuniformities. For example, a 1.5-μm-thick layer of AZ-1350J photoresist was deposited on a silicon surface with microscopic roughness of amplitude 0.15 μm. Irradiation of rotating samples with a 2 keV, 20 μA/cm^2 argon beam at angles of 0, 40, 60, and 70° showed that in the first case the photoresist was completely sputtered, but the roughness was not smoothed out. At 60° the maximum surface deviation was reduced to 0.1 μm and, over a narrow region (0.2 μm), to 50 nm [196]. This technique has been used to obtain 50-mm-diameter disks of $LuYSm_{2.2}(CaGe)_{1.6}Fe_{4.2}O_{12}$ with a spacing of 8 μm that contain 68 kbits of cylindrical magnetic domains [165]. Ion implantation was used to eliminate hard domains.

The rate of ion etching of garnets is usually 0.05-0.1 μm/h. Experiments have shown that radiation damage from preliminary ion bombardment causes an increase in the rate of ion and chemical etching [197]. The rate of etching of $YGdTmFe_{4.2}Ga_{0.8}O_{12}$ garnet by phosphoric acid was increased by factors of 2, 40, and 1000 compared to that of unirradiated samples following bombardment by H^+, He^+, and Ne^+ ions to a dose of $1 \cdot 10^{16}$ cm^{-2} at energies of 50, 100, and 300 keV, respectively. Calculations of the nuclear slowing down for these ions at the maxima yield the proportion H:He:Ne = 1:10:300, which is in good agreement with the enhanced etching observed in the irradiated samples.

Ion milling has been used for making permalloy circuits with 1- and 2-μm-diameter cylindrical magnetic domains with gaps of 0.25-0.5 μm between neighboring elements, so that information could be recorded with a density of over 106 bits/cm^2 [198].

Ion-beam micromachining of optical components for integrated circuits. A duopigatron ion source, holographic techniques, and scanning electron beam lithography have been used to obtain gratings with a line thickness of 100 nm and a spacing of 280 nm in an optical waveguide in 10-μm-wide, 2-μm-thick ZnS with an etching depth of 50 nm [199]. The following components of optical circuits have also been fabricated: linear and curved waveguides, directional couplers, thin-film analogs of lenses and prisms, modulators, infrared polarizers, acoustic wave devices, silicon and gallium arsenide integrated circuits, sensitive structures for surface wave reflectors, etc. [199, 161]. Ion micromachining of integrated optical circuits is a controllable and reproducible process. The image quality and characteristics of optoelectronic devices are limited only by the quality of the photographic masks and the optical copying process.

Manufacture of Josephson junctions. Ion beams have been used for manufacturing Josephson junctions in 50-nm-thick niobium films [200]. Slits of width 0.6 and depth 250 μm, as well as junctions of thickness 0.6 μm, were obtained.

Ion bombardment changes the critical temperature for the transition into the superconducting state. Depending on the irradiation dose it may vary from 8.5 to 2.5 K.

Thinning of samples for electron microscopy and removal of layers for analysis of the depth distributions of elements. Ion micromachining is widely used for thinning of samples for direct examination by electron microscopes, Auger analyzers, secondary ion mass spectrometers, and other instruments [195]. It can also be used to prepare very thin samples (down to 10 nm) that are almost free of damage and contamination, to study the depth distribution of implanted ions, and to study oxidation and changes in the elemental composition of near-surface layers. As an example, the CAMECA ISM5300 microanalyzer has been used to examine phenomena that occur during the interaction of 1-15 keV $^{16}O^+$, $^{16}O^-$, and $^{40}Ar^+$ with clean monocrystalline silicon: the effect of adsorbed and incident oxygen on the secondary ion emission coefficient and the dynamics of the implantation zone. For argon ions with an energy of 1 keV at a current density of 15-30 $\mu A/cm^2$, the sputtering rate was 0.8-1.8 nm/min.

Sharpening field emission sources. By using 1.5-2 keV neon ions and annealing tungsten field emission sources at 1000 K for 10 s, it was possible to obtain points with a radius of curvature less than 20 nm. The conditions for and results of this process are shown in Table 5.12. The rate of reduction in the radius of curvature is proportional to the neon pressure:

$$-dr/dt = kP_{Ne},\qquad(5.10)$$

where $k = 3.5 \cdot 10^3$ nm·Pa/min at an ion current of $1 \cdot 10^{-5}$ A.

Shaping optical surfaces by small cross-section ion beams. Methods have now been developed for producing various surface profiles (parabolic, wedge shaped, spherical). Intense ion beams and plasmas are available for this purpose, and the technical feasibility of and need for ion machining of optical lenses, mirrors, patterns, etc., has been demonstrated [202-204]. Mirror shaping involves the greatest difficulty because of the large size of mirrors and the importance of surface deviations (4 times more critical than for lenses).

Studies have been made of the ion polishing of transparent fused quartz (TVS), glasses with ultralow expansion coefficients (ULE), and Cer-Vit (C-101). Beams of $^{40}Ar^+$ ions were accelerated in a 300-keV Cockroft–Walton accelerator made by Texas Nuclear, Inc. At these energies secondary electron emission may cause large errors in measurements of the sputtering yield (Table 5.13). Another limiting factor is the divergence angle of the beam.

The forward scattering fraction was measured to be 0.00063 after ion polishing, which is comparable to a very good standard polish (0.00060). The central region (12 cm diameter) of a piece of TVS quartz was irradiated by 150-keV argon ions with an average beam current of 120 μA. The initial root-mean-square deviation was 0.05λ, which fell to 0.01λ after processing. The experiments were long, however, and required 200 h of irradiation at an ion current of 150 μA. At minimal voltages the ion current can be raised to 20 mA in pulses. Removal of a 250-nm-thick ($\lambda/2$) layer of glass in a three-meter telescope requires 30 days of processing. This type of telescope has been proposed for the orbital astronomical observatory.

Fabrication of nuclear filters and ultrafine metal fibers. Fast heavy ions passing through a film of mica, glass, or Plexiglas form channels with high radiation damage. In these channels complex molecules are broken up into smaller components. The intermolecular binding energy is reduced. This causes faster

TABLE 5.12. Conditions for and Rates of Sharpening of Tungsten and Field Emission Emitters at a Neon Pressure of $6.7 \cdot 10^{-3}$ Pa and Ion Current of $1 \cdot 10^{-5}$ A [201]

Duration of sputtering (min)	Voltage during ion bombardment at a current of $1 \cdot 10^{-5}$ A (V)	Voltage after ion bombardment with an extracted current of $1 \cdot 10^{-6}$ A (V)	Tip radius (nm)	Rate of sharpening (nm/min)
0	1305	1305	60.0	1.4
5	1128	1111	53.0	2.4
10	922	903	41.2	2.4
15	702	672	29.0	2.4

TABLE 5.13. Sputtering Yields of Optical Glasses for 150-keV $^{40}Ar^+$ Ions [203]

Glass	ρ (g/cm^2)	M	Number of atoms/molecule	Sputtering yield (atoms/ion)
TVS	2.2	60	3	1.9
UL E	2.2	61.4	3	1.86
Cer-Vit*	2.51	252	14	3.7

*$Li_2O \cdot Al_2O_3 \cdot 2SiO_2$ bombarded by 80-keV $^{126}Xe^+$ ions.

chemical etching, as noted above. Etching in thin films forms pores whose diameters may range from 3 nm to several tens of microns [116]. Various materials have minimum (threshold) values of the specific ion energy loss at which selective etching is possible. As an example, for Plexiglas it is 3-3.5 MeV·cm^2/mg. The ionization losses increase rapidly with the charge of the bombarding ion. For He$^+$, $^{12}C^+$, $^{20}Ne^+$, and $^{40}Ar^+$ ions the peak total ionization loss lies in the range of 0.2-1 MeV/nucleon.

Microfilters with channel radii of 12-50 nm and 10^7-10^9 holes/cm^2 have been made by bombarding 3-10-μm-thick Plexiglas with 45-MeV Ar^{8+} ions. These microfilters can sustain a pressure of up to $3 \cdot 10^5$ Pa. The capacity of these filters in water ranges from 0.0055 to 0.015 cm^3/(cm^2·atm·min).

It should be noted that radiation porosity is observed during irradiation of metals by their own ions. For example, pores with sizes from 2 to 160 nm have been obtained during bombardment of 0.15-mm-thick commercially pure nickel, which had initially been vacuum annealed at 800°C for 1 h, by 45-keV nickel ions at a current density of 3 μA/cm^2 to a dose of $1.6 \cdot 10^{17}$ cm^{-2}, corresponding to roughly 40 displacements per atom, at a temperature of 400-700°C [205]. The corresponding pore densities ranged from $4.5 \cdot 10^{17}$ to $5.75 \cdot 10^{12}$ cm^{-3}.

Nuclear molecular-virus filters can be used for cold stabilization of beer, wine, and other liquid food products to preserve them for prolonged periods at room temperature. Nucleopores can also be employed for sterilization of biological and microbiological media, for separating different types of cells (especially isolating cancer cells in blood), and for removal of bacteria from water.

Channels of atomic size (2-20 nm) have been obtained in dielectric matrices by using a 1-GeV proton beam at a dose of $2 \cdot 10^{14}$ cm^{-2} [206]. Porous matrices were produced in mica and fianite (ZrO_2). Ultrathin metal wires with diameters of 5-10 nm have been formed by pressing a liquid metal into the etched channels. Ultrathin metal wires of this sort in dielectrics have superconducting transition characteristics that open the possibility of raising the critical temperature of superconductors. Mercury wires with these diameters in mica have $T_K = 3.5\text{-}4$ K.

The average intensity of the 150-MeV ion beam at the U-300 accelerator at Dubna is $5 \cdot 10^{12}$ s^{-1}, which makes it possible to irradiate hundreds or thousands of square meters of film per day. Widespread use of nuclear sieves will make it possible to proceed to a qualitatively higher level of productivity in many branches of the national economy.

Other applications of ion processing of materials. *Revealing stressed states in metals by ion bombardment* [207]. Samples of copper, duraluminum, and steels (St. 3, steel 20, steel 45, steel 40Kh, steel U8) with Brinell hardnesses from $27 \cdot 10^7$ to $550 \cdot 10^7$ Pa were stamped with the number "2" in a hydraulic press at a force of 2940-19,600 N. After a deformed region had been created by the press, the surface layer of the metal was mechanically ground and polished until the number "2" disappeared completely and the sample was then bombarded with 4-keV (in a glow discharge at a current density of 1.5-2 mA/cm^2) and 40-keV (in a magnetic separator at 0.4 mA/cm^2) Ar$^+$ ions. Bombardment was continued for from 15 min to 2-2.5 h or when the number appeared. This operation was repeated until the contrast of the deformed zone disappeared completely. The maximum depth at which it was still possible to distinguish a number was as much as 1-1.5 mm in a number of cases (for steel types 45, 20, and St. 3).

The effect by which the contrast of images is revealed, including by bombardment with a 37-MeV proton beam, is explained by different grain sizes in the deformed and undeformed regions and by radiation-stimulated diffusion of defects and impurities in the mechanical stress field, as well as by a difference in the sputtering yield for these regions.

Fabrication of microscopic cutting tools (microdies, chisels, and bits) with dimensions below 50 nm [208]. Ion bombardment of different facets of monocrystals creates oriented faceted figures in them: three-, four-, and six-faceted pyramids with the symmetry of the facet on which they develop [62]. Cutting tools with a hardness comparable to that of a steel file have been made by filling the resulting pits with dust of nickel+phosphorus and nickel+boron alloys.

Studies of the mechanism for corrosion by means of ion thinning of oxides [209, 210]. The Mark III ion micromachining instrument (IMMI Mark III) has been used in transillumination electron microscope studies of the corrosion of zircalloy-4, which is used as cladding for the fuel elements of nuclear reactors. Two ion beams were used simultaneously to remove material from both sides of the oxide. The beams were operated with an accelerating voltage of 4-6 kV, an ion current of 100 μA, and an angle of incidence of 15°. The rate of removal of the oxide was roughly 1.5 μm/h, and the thickness of the oxide films was 1-6 μm.

A CAMECA ion analyzer has been used to study the mechanism for corrosion of a tantalum alloy in lithium in a heat pipe operating at a temperature of 1600°C with a heat flux of 157-170 W/cm^2. An rf ion gun produced a beam of argon ions with an energy of about 15 keV. The sputtered atoms, part of which escaped as ions, provided an image of the surface in O^-, O^+, Y^+, Li^+, and Ta^+ ions with the aid of an immersion objective. It was suggested that corrosion occurs as a result of transport and localization of oxygen in the heating zone.

Chapter 6

ION FINISHING AND EXPOSURE OF
STRUCTURE IN SOLID SURFACES

6.1. Ion Grinding and Polishing of Metal and Glass Surfaces

Seventy percent of all artificial diamonds are used in machine building and metalworking for finishing processes. Raising the surface quality of chromed hydraulic cylinders from roughness class 8 to class 10 increases their wear resistance by a factor of 4 [211].

However, mechanical grinding and polishing have important disadvantages. First, abrasive materials contaminate the worked surface. This leads, for example, to a deterioration in the resistance of dielectric mirrors to damage by laser radiation which is absorbed by micron-sized inhomogeneities. Second, the surface layer is destroyed to a depth of as much as 200 μm. The optical systems of CO_2 gas lasers and optoacoustic waveguides place exceptionally high demands on surface quality, which is not determined solely by the final polishing operation. Flow lines, or traces of plastic deformation of KRS-5 and KRS-6, have been observed to a depth of 100 μm. Third, the lifetime of diamond grinding and polishing wheels in the assembly lines for automatic machining of optical surfaces in the GK-44 objective is not sufficient. Automatic replacement of the instrument is necessary.

Ion-beam grinding and polishing eliminate these shortcomings of mechanical processing. Ion-beam processing has been used for polishing and activation of substrates for optical coatings [212].

The thin films prepared by vacuum deposition have a microscopic surface roughness of 5 nm to 0.5 μm. Films of the SiO_2–MgO oxide mix have been obtained with the URM.3.279.011 system by electron-beam evaporation and polishing in a single cycle with a 440-kHz rf discharge. The surface quality was evaluated before and after ion polishing using microphotographs with a magnification of 20,000× and by measuring the attenuation of He–Ne laser light in the film. Before the 630-nm-thick film was polished, the surface nonuniformities were of size 30-60 nm. Polishing by 3-keV oxygen ions for 5 min reduced this size to 10 nm, and the attenuation in the film after ion polishing was reduced by an order of magnitude (from 50 to 5 cm^{-1}). Replacing the oxygen ions by argon lowers the surface quality of the SiO_2–MgO films. A substantial number of regular formations with sizes of about 100 nm appear against a general background with microscopic irregularities of less than 15 nm.

Similar results have been obtained for films of ZrO_2 and SiO_2. Prior to ion polishing the sizes of the microscopic irregularities were 15-30 nm with individual particles up to 100 nm. After polishing of the films by 1.5-keV argon ions for 20 min the size of the microscopic irregularities was reduced to 5 nm. The size of the crystalline grains in a 0.7-µm-thick ZrO_2 film was 50-80 nm, and after polishing by 1.6-keV argon ions for 18 min the surface was considerably smoother, with a background level of less than 15 nm.

Films of SiO_2–Ta_2O_5 and Ta_2O_5 with thicknesses of 0.1-0.2 µm produced by rf ion-plasma sputtering had a general background of 7-10 nm particles with a considerable number of formations with sizes of 25-60 nm, and massive coagulations of sizes of up to 200 nm. After polishing with 3-keV oxygen ions for 5 min, the size of the crystalline grains was reduced to 5 nm, the film surface was uniform, and the light scattering coefficient fell from 0.015 to 0.005%.

A nine-layer mirror in which each layer had been ion polished after deposition (with 25-35 nm removed) scattered 0.01% of the light, i.e., the same as the substrate.

K8 optical glass or quartz has isolated microscopic irregularities of size 7-10 nm after ordinary mechanical polishing prior to deposition of metallic coatings on the surface. Polishing for 15 min with 1-1.5 keV argon ions at a current density of 1 mA/cm^2 and a sputtering rate of 1 µm/h (0.25 µm of glass was removed) greatly improves the surface quality of glass ($R_Z < 5$ nm).

Ion sputtering activates a glass surface, creating free surface bonds which become artificial nucleation centers during deposition. After 2 s of deposition the surface is already completely covered with small nuclei. After 60 s a continuous film with a thickness of 5.5-8 nm is formed. A continuous film does not form in this time over a surface that has not been irradiated by ions. Extremely uniform (in terms of shape and nucleus size) films of aluminum, copper, chromium, etc., are formed on ion-polished glass surfaces, even for low deposition rates. This emphasizes the paramount role of ion polishing in the initial stage of layer growth.

Similar results have been obtained for different regimes of thermal and ion-plasma vacuum deposition of metal films. This indicates that the mechanism of nucleation on atoms with free bonds is general in nature. The surface is activated after 3 min of ion sputtering and 50 nm of glass has been removed. This fact is of great practical importance, since it considerably reduces the time required to clean a surface. Keeping substrates in the vacuum for 30-60 min after ion polishing reduces the degree of activation by 30-40% because of the interaction of the activated sur-

TABLE 6.1. Etching Rates for Different Materials by Argon Plasma Flows with Energies of 170 eV (data of Nikonenko and Trofimov [144])

Material	Etching rate		Sputtering yield, atoms/ion
	nm/min	rel. units	
Al	13	1.1	0.47
Si (100)	15	1.25	0.45
Si (polycrystalline)	19	1.6	0.68
Ti	7	0.6	0.24
V	7	0.6	0.31
Ni, Fe	12	1.0	0.65
Mo	9	0.75	0.34
Ta	6	0.5	0.20
SiO_2	17	1.4	0.22
FP-383	15	1.25	–
AZ-1350	18	1.5	–
ÉLN-200	30	2.5	–

face with residual gas molecules. Thus, the substrates must be bombarded again for 3 min before a new layer is deposited.

Ion cleaning, polishing, and activation of surfaces increases the adhesion of metallic and dielectric coatings to substrates of glass, metal, and acrylic plastic. This improvement in adhesion has been evaluated by stripping and by examining the quality of separation when the films are cut on engraving-ruling machines. Metal coatings on optical glasses are used in the fabrication of scales, grids, and screens.

We now estimate the time required for ion polishing. According to the GOST 2789-73 standard, the value of the roughness parameter R_a lies within the interval 0.008-100 μm, and the maximum height of profile irregularities is $R_{max} = 0.025$-1600 μm.

Using Eq. (4.1) for the sputtering rate and taking the sputtering yield of iron to be 2 atoms/ion, the ion current density at the sample to be 0.1 A/cm^2, $M = 56$, and $\rho = 7.8$ g/cm^3, we find the dependence of the maximum sputtering time on the difference between the least and greatest heights of profile irregularities prescribed in the GOST standard to be

$$t_{max} = \frac{\Delta R_{max}}{v_s} = \frac{0.16 \cdot 7.8}{10^{-5} \cdot 10^{-1} \cdot 2 \cdot 56} = 1.1 \cdot 10^4 \text{ s},$$

or about 3 h. If the height of the surface irregularities must be reduced from 2.5 to 0.16 μm (i.e., the surface quality class raised from from 6 to 9), then ion polishing can do this operation in only 16 s. These estimates agree with the above experimental data on polishing of optical glasses.

It has been shown that a metal surface can be polished with an ion beam [213]. The surface of a copper disk machined on a lathe had an average microscopic roughness height of 0.5 μm. An argon beam with an energy of about 8 keV, a current of 1 mA, and a diameter of 1 cm at the target was aimed at the sample at an angle of 85-88° to the surface normal. The sample was rotated about the normal at 5

TABLE 6.2. Optimum Conditions for Etching Various Solids by Ion Bombardment [62, 216]

Substance	Ion (gas)	Pressure, Pa	U, kV	j, mA/cm² (I, mA)	τ, h	Exposed feature
Metals and alloys						
Ta, Ni, Mo, W	Ne+	67-80	3-5	7	3	Crystal structure
Al, Ni	Ar+, Ne+	1.3-13	1.5-3	1-1.5	0.5-1	"
YalT Steel	Air	13	6	10	0.5	Austenite and ferrite
Steel with 0.45% C	Ne+	0.13-13	1.7	0.4	1.5	Perlite and ferrite
St. 1 (not subject to chemical etching)	Ne+	2.7	1.7	0.3	2	Martensite
Bronzes: 77-80% Cu, 1-2% Pb, remainder Zn	Ar+	2-2.6	2-5	0.6	0.12	Optical and electron microscopic structure
Cu	Ar+	2-2.6	2-5	0.6	0.17	"
Ni	Ar+	2-2.6	2-5	0.6	0.25	"
Fe	Ar+	2-2.6	2-5	0.6	0.49	"
U	Ar+, Kr+	0.01	5	(1-3)	-	"
Dielectrics						
Quartz	Ne+	11	2.3-9	1.5	12-54	Crystalline structure
Amber, plexiglas	Ne+	11	1.5	2-2.5	2-3	"
Forsterite ceramic	Ne+	13-7	1.2-2.8	0.2-1	5	"
Aluminum oxide ceramic	Ne+	7-13	2.7-3.2	0.2-2.5	3-5	"
Oxides of uranium and aluminum, carbides of nickel, titanium, and cobalt, titanium silicide, titanium diboride, porcelain	Kr+	5	4	(2)	0.1-0.5	"
Semiconductors						
Ge(100), Si(111)	Ar+(?)	0.1	0.5-0.1	0.1	6-17	Etching and dislocation patterns
Synthetic materials						
Nylon 66, polyethylene, and terephthalate	Air	665-3325	4	(30)	0.03-0.08	Crystallite period

rpm. The background pressure of commercial argon was maintained at $6.5 \cdot 10^{-2}$ Pa. After a dose of $7.5 \cdot 10^{19}$ ions/cm^2 had been reached, the height of the roughness was reduced to 0.05 μm. Under these ion bombardment conditions, exposure of and changes in the structure of the copper were not observed.

Studies have been made [214] of ion-beam polishing of type 4604 molybdenum alloy, which is used as an electrode material in the ion-optical systems of sources for megawatt hydrogen ion beams in thermonuclear experiments. A cold cathode Penning ion source was used. The ion energy was varied over the range 50-1500 eV and the current density, over 0.02-1 mA/cm^2. The temperature of the samples was kept at 400 K. The rate of abrasion by 200-eV argon ions at a current density of 80 μA/cm^2 was 10-15 nm/min. It increased by a factor of 3-10 when the chemically active gas CF$_4$ was used. Adding 15 vol. % of oxygen or 30 vol. % of argon to the CF$_4$ sharply increased the abrasion rate. For an ion energy of 1500 eV and current density of 1 mA/cm^2 the time required for ion-chemical polishing of 4604 alloy from surface finish class 6 to 10 was 50 min.

Calculations show that smoothing of microscopic surface irregularities probably will occur at low ion energies, down to some energy below the sputtering threshold, without having the ions strike the surface at an oblique angle and without rotating the sample [144, 215]. At low ion energies, radiation damage of the layers near the surface is greatly reduced, but the polishing process takes longer. When a plasma accelerator, such as a Penning discharge, with an average argon ion energy of 170 eV was used with normal incidence of the ions, fairly high etching rates were obtained for different materials at a discharge voltage of 250 V and currents of 1-2 A (Table 6.1). When xenon is used, the etching rates shown in Table 6.1 are raised by factors of 1.3-1.5.

6.2. Exposure of the Surface Structure of Solids by Ion Bombardment

Sputtering by ion bombardment is a unique method for exposing the grain, block, and dislocation structure of metals, alloys, semiconductors, minerals, polymers, and synthetic and biological materials [62].

Ion etching has a number of important advantages over chemical and thermal methods: (1) It can be used to expose the structure of the most durable and hardest materials, which are not subject to other etching techniques. (2) Ion bombardment is a universal etching agent, suitable for any solid. (3) During ion etching oxide layers are not formed, the sample is not contaminated by the etching, abrasion, or polishing materials, and the surface structure is not destroyed. (4) The ion etching depth can be controlled. (5) It is possible to collect the etched material in order to analyze it later. (6) There is no difficulty in etching radioactive or chemically active substances. (7) Ion etching easily exposes the structure of metals heated to high temperatures.

Ion etching is done in a glow discharge or with an ion beam. Special equipment has been developed and manufactured by industry for ion etching, including the following systems: UIT-4, UVR, and ITR. Inert gases and, sometimes, air are used as working gases. The ion energy is varied over the range 0.3-10 keV, the current density over 0.1-20 mA/cm^2, and the irradiation time over 1-60 min or more. Grain boundaries and structure are exposed primarily because of the differences in the sputtering rates for atoms with different binding energies in the solid surface.

Table 6.2 shows the conditions for etching various substances.

The average depth of etching of the boundary between grains of commercial copper for a constant accelerating voltage of 2.3 kV and neon pressure of 5.7 Pa was 0.15 μm after 0.5 h and reached saturation (0.57 μm) after 2 h. As the accelerating voltage was raised, the etching depth increased almost linearly [62].

6.3. Topography and Morphology of Surfaces Created by Ion Bombardment

The surface topography of solids changes under ion bombardment. Spivak et al. [62] have shown that these changes take place over time in several stages. At first the surface is cleaned of contaminants. Then crystalline structures are formed inside the grains. At the same time, cones develop and are destroyed. Often blisters are formed on the surface and it undergoes flaking.

Craters, cones, facets, crests, pyramids, amphitheaters, mushrooms, spongy structures, blisters, and scales are formed on the surface of metals by ion bombardment [217, 218]. The formation of these structures is determined by many factors: general and selective sputtering yield, the dependence of the sputtering yield on the angle of incidence of ions on elements, the initial surface nonuniformity, the concentrations of atoms, the ion energy and dose, the binding energy of atoms, the angular distribution of sputtered atoms, the surface temperature, the energy flux at the surface, the presence of and degree of contamination of the surface by weakly sputtered atoms, the crystal structure and orientation of crystals and monocrystals, focusing of atomic collisions, etc. [219].

As an example, Tanovic et al. have made a detailed study [150] of sputtering of high-purity copper (99.999%) with large facets by 40-keV argon ions at a dose of 2·10^{19} cm^{-2} which showed that etching pits are formed on almost all cleavage planes and cones are formed only on some of them – those with an axial direction along <1131>. The cones have such regular shapes, sizes, and mutual separations that daylight is selectively absorbed by them and only reddish light is reflected.

Several theories and computer programs have been developed to follow the development of microreliefs in the form of cones and triangular facets [100, 165, 170, 219]. The size of these formations may reach 10 μm. It has been shown [100] that cones can grow on seed inhomogeneities smaller than the length of a collision cascade only within a relatively narrow temperature range (around 200°C), $T_1 < T < T_2$ with $\Delta T = T_2 - T_1 = (1/30\text{-}1/50)U_0$, where U_0 is the sublimation energy. This temperature interval can be found using the equations

$$T_1 = U_0 \Bigg/ \left(40 \ln \left[\frac{(0.017 \, a^2 \, v)^{1/4}}{j \, Y} \Big/ L \right] \right) \qquad (6.1)$$

and

$$T_2 = 0.15 \, U_0 / \ln (L/0.2a), \qquad (6.2)$$

where a is the size of an atom, v is the vibrational frequency of the atoms (on the order of 10^{12} s^{-1}), j is the current density of the bombarding particles (on the order of 10^{15} cm$^{-2 \cdot}$s^{-1}), Y is the sputtering yield (atoms/ion), and L is the size of a collision cascade (cm). For light ions with energies of 1-100 keV, $L \sim 10^{-6}$-10^{-4} cm [103].

The characteristic temperature at which a rapid, exponential growth in the sputtering yield begins is given by

$$T^* = 0.7 \, U_0 / \ln (2 \, v\tau), \qquad (6.3)$$

where $\tau = (jYa^2)^{-1}$ is the time for a single atomic layer to undergo sputtering. The characteristic temperature for the transition from the collisional to the evaporative mechanism for cathode sputtering is

$$T^* \approx U_0 / 40. \qquad (6.4)$$

Some questions may arise in connection with the above remarks: how does ion polishing of the surface occur, and are the methods for measuring surface roughness sufficiently accurate?

We now supplement the discussion of Section 4.4 with the following additional data. Extremely sensitive measurement techniques have been used in studies of surface roughness: interferometers with an error of less than 0.0008λ at the wavelength of a helium–neon laser ($\lambda = 632.8$ nm) [203], electron microscopic studies with a magnification of 20,000×, and measurements of the attenuation of helium–neon laser light [212]. A profilometer with a height sensitivity of ± 2.5 nm at a magnification of $10^6\times$ has also been used [220]. The limiting sensitivity of the Dektak profilograph, which has a microcomputer, is as good as 0.5 nm. The sputtering of sapphire in three orientations (1120), (1021), and (0001) by normally incident 30-keV argon ions to a dose of 0.00107-0.0305 C/cm^2 over 30 min has been studied. The average total dose was 0.2263 C/cm^2. Under these conditions a small increase in the microscopic surface irregularities was observed with increasing dose for the (1120) plane and a small reduction, for the (1012) plane.

The feasibility of ion polishing of diamond as the holder is rotated at an angle of 65° to the normal has been studied in terms of the reduction and disappearance of triangular microscopic inhomogeneities [221]. A drilling rate in diamond of 130 μm/h was obtained with a 7.5-keV, 100-μA argon ion beam. Conical holes with diameters from 0.05 to 2.26 mm were drilled through 1.58-mm-thick diamonds.

McNeil and Herrman [164] have made detailed studies of changes in the microscopic inhomogeneities on copper, molybdenum, silicon, nickel deposited on aluminum, glass, ThF_4, and ZnS during bombardment by 500-1500 eV argon ions at 60° to the normal with an overall current of 50-100 mA and a current density of 0.5-1.0 mA/cm^2 [164]. A small increase in the inhomogeneities (by 0.5 nm) was observed in electrolytic nickel, silicon, quartz, and BK7 glass after removal of a 317-nm layer. More growth was observed on nickel, molybdenum, and ThF_4; however, in these experiments the samples were not rotated.

Ion-beam sputter deposition of dielectric coatings reduces the roughness of copper optical surfaces, which varies over 1.6-12 nm. The following sputtering rates (nm/s) were observed with 500-eV (1500-eV) ions at a current of 1 mA/cm^2: ZnS, 0.032 (0.081); ZnSe, 0.047; ThF_4, 1.81; MgF, 0.79; and BK7, 0.55 (2.31).

Chapter 7

DEPOSITION OF FILMS
BY ION SPUTTERING

7.1. Thin Film Coatings and
Methods of Producing Them

Certain properties of many structural steels and cast irons are not satisfactory for the conditions under which they are used. In order to raise the corrosion resistance or wear resistance or to reduce friction at their surfaces, these metals are coated with films of elements to improve their properties. Another application for ion sputtering is the production of thin films of ferromagnetic materials (thickness 0.01-$10\,\mu$m) for use as memory elements in computer technology. Magnetic thin films of permalloy can be remagnetized in 10^{-9} s. The density of information recorded on them can be as high as 100 bits/mm^2 [222].

By means of cathode sputtering of niobium and germanium, in 1973 Cavaler [223] first obtained a Nb$_3$Ge superconductor with a record-high transition temperature (23 K) that permitted the use of liquid hydrogen in place of the more expensive liquid helium. By simultaneous or sequential sputtering of two or more elements it is possible to obtain multicomponent or multilayer junctions containing a wide range of chemical compounds and mixtures.

Particular success has been achieved in the commercial utilization of ion-plasma technology for microelectronics. Ion-plasma surface cleaning, plasma anodizing of silicon wafers, ion, plasma-chemical, and ion-chemical etching of microstructures, plasma-chemical stripping of electron-resistive masks, ion implantation, plasma-chemical deposition of dielectric films, the growth of monocrystalline films, and the fabrication of metal contacts and junctions by cathode sputtering are all engineering processes, which with computer control have made possible the transition from liquid chemical to "dry" technology [104-108].

Semiconductor integrated circuits based on a planar process were first created in 1959-1961, but by 1972 the world industry was already producing more than 1 billion integrated circuits a year [224]. At the end of the 1970s the production of integrated circuits and discrete devices had reached a level of 10-12 billion pieces per year [105]. They are used in applications ranging from computer and space technology to household appliances. A distinction is made between thick film technology for the manufacture of integrated circuits, where the layers of conducting, resistive, and dielectric material are 1 to 25 μm thick, and thin film technology, where the thicknesses are up to 1 μm.

Microelectronics continues to develop both in terms of increasing the packing density in integrated circuits and the level of integration and in terms of creating functional electronics. The first aspect has led to levels of integration characterized by tens of thousands of elements in a single integrated circuit package with the individual elements having micron and submicron dimensions. One task is to raise the level of integration in microcircuits to 10^6 elements on a crystal [105]. The development of materials, techniques, and technology for fabrication on a molecular level, i.e., molecular electronics, has begun. The density of elements may reach fantastic levels, 10^{15}-10^{18} cm^{-3} [225]. The second aspect of the development of microelectronics involves a search for new physical phenomena and principles for fabricating electronic devices with specific functions at the circuit and system levels. These include optical phenomena in solids (optoelectronics), the interaction of fluxes of electrons with sound waves in solids (acoustoelectronics), superconductivity, the properties of magnetic substances and semiconductors in magnetic semiconductors (magnetoelectronics), and other phenomena.

The fabrication of microelectronic devices has required the development of the physical, chemical, technical, and engineering foundations for a most massive and complex production effort. The physical and chemical processes in microelectronics technology have been examined in detail elsewhere [105]. The techniques for depositing thin film coatings [226] differ according to the methods of obtaining a flux of material [227]: in neutral or partially ionized states or in the form of an ion beam; evaporation from a heated crucible, cathode sputtering in a discharge, or ion-beam sputtering into vacuum; evaporation by laser beam, electron beam, vacuum arc, or autonomous ion sources. The methods of thin film deposition differ in the mechanism of particle transport: through a plasma or a neutral reactive medium; through a vacuum with ionized vapor, gas, or ions that may or may not have been separated in mass; with or without chemical reactions involving the working gas.

Finally, thin film deposition techniques differ in the conditions under which the material is condensed on the substrate: without bombardment of the film by fast particles; with bombardment of the film by particles of the working gas; with bombardment of the film by vapor particles; with and without application of a bias voltage to the substrate.

We now examine some of the latest advances in thin film production by means of cathode and ion-beam sputtering, ion deposition, and deposition from ion beams.

7.2. Production of Diamond
and Diamondlike Films

Method for producing films of artificial diamond using ion sputtering of graphite [130, 228]. This method, proposed in 1971 by V.M. Golyanov and A.P. Demidov, is based on cathode sputtering of graphite in a magnetic field with an inert gas pressure of 10^{-5}-10^{-2} Pa and the deposition, from at least two sputtering sources, of neutral carbon atoms on a solid substrate cooled to 100 K. The physical idea is to use relatively fast "active" sputtered atoms with energies low enough to prevent radiation damage of the diamond films that result from a suitable geometrical orientation of the condensing atomic beams.

Films of artificial diamond are obtained on an ion sputtering apparatus [229]. In order to improve the quality of the deposited layers, the discharge chamber is made up of several sections, each of which has four disk cathodes whose centers are located at the vertices of a nominal lattice, for example, an octahedron. The cathodes are attached to a cooled base and the anodes, to a coiled pipe with coolant. The process is further automated by equipping the crystal holders with manipulators and devices to count the number of layers, and an electromagnetic valve is placed in the inert gas supply system and connected electrically to the power supply circuit. The system requires a power of 2 kW for sputtering. The growth rate of the artificial diamond layers is 0.8 nm/min for a gas discharge voltage of 4 kV and current of 4 mA. The apparatus weighs 150 kg, and its dimensions are $0.8 \times 0.8 \times 2$ m^3. A peak power of 4 kW is required during outgassing. The artificial diamond films have roughly 20 times the tensile strength of natural diamonds. They are also harder than natural diamonds.

The basic characteristics of the artificial diamonds are [230]:

Hardness on Mohs scale	≥ 10
Tensile strength	$6.9\text{-}7.8 \cdot 10^9$ N/m^2
Modulus of elasticity	$1.47 \cdot 10^{12}$ N/m^2
Density	3.4-3.7 g/cm^3
Electrical breakdown strength	$2 \cdot 10^6$ V/cm
Specific resistance of the unalloyed material at room temperature	$5 \cdot 10^6$ $\Omega \cdot$cm
Thermal coefficient of resistance	Negative
Type of conduction with boron doping	p-type
Type of conduction with nitrogen, phosphorus, or bismuth doping	n-type
Chemical stability in boiling HNO$_3$, aqua regia, or chromium mixture	Does not react

TABLE 7.1. Parameters of the Sputtering and Deposition Processes for Production of Carbon Films with Diamondlike Properties with Different Ion Sources [228]

Parameter	Sputtering of graphite			
	Penning cell	duoplasmatron	vacuum arc	Finkelshtein source
Energy of deposited particles, keV	0.001–0.1	0.01–0.04	0.03	0.5–10
Gas pressure in ion source, Pa	$10^{-5}-10^{-2}$	0.3–6.7	–	0.013–0.13
Gas pressure in working vessel, Pa	$10^{-5}-10^{-2}$	$6.7 \cdot 10^{-7}-10^{-4}$	$1.3 \cdot 10^{-4}$	$1.3 \cdot 10^{-4}$
Deposition rate, nm/s	0.013	0.3–1.5	0.5–1.0	0.3–1.7
Substrate temperature, K	77	300	–	375
Substrate dimensions, mm	22	100	200	30–100

Parameter	Destruction of hydrocarbons		
	glow discharge	PIG discharge	magnetron source
Energy of deposited particles, keV	–	0.2–0.8	–
Gas pressure in ion source, Pa	0.13–1.3	0.13–1.3	0.67
Gas pressure in working vessel, Pa	0.13–1.3	$6.7 \cdot 10^{-2}$	$1.3 \cdot 10^{-2}$
Deposition rate, nm/s	0.2–0.8	0.5–1.2	12.2
Substrate temperature, K	77–343	300	60–327
Substrate dimensions, mm	50	–	60

Thin films of artificial diamond have been used to prepare superconducting films of technetium, tin, molybdenum, and vanadium with enhanced critical parameters [231]. They can be used in electronics for the fabrication of passive and active components, for protecting materials from corrosive media, and in the jewelry industry.

In order to obtain films of artificial diamond the discharge chamber is outgassed for 10 h at a temperature of 250°C. The residual pressure of hydrogen must not exceed 10^{-7} Pa, that of nitrogen, oxygen, and water vapor, 10^{-8} Pa, and that of the working gas, krypton, $6.5 \cdot 10^{-4}$ Pa. The thickness of artificial diamond films for protecting superconducting metal films is 1.5-4.0 nm. The rates of deposition of carbon, molybdenum, and niobium are 0.27, 0.45-2.45, and 5 nm/min, respectively, depending on the discharge current.

Methods of producing diamondlike films. Diamondlike films are obtained in various ways: from the gaseous phase by growth on the facets of natural diamond [232]; deposition of fast ions [233, 234]; sputter deposition by ion beams; vacuum evaporation; rf sputtering using an inert gas and hydrocarbon gases; dc glow discharge with hydrocarbon gases and a small admixture of argon; pulsed coaxial plasma guns using methane; vacuum arc with a graphite cathode; deposition of an ion beam with argon ions and hydrocarbon fragments; and deposition using carbon ion beams, as well as combinations of these methods [235].

TABLE 7.2. Characteristics of Diamond, Diamondlike, Carbon-Containing, and Graphite Films

Film	Specific resistance ($\Omega \cdot cm$)	Microhardness (Vickers)	Refractive index	Color
Diamond (for comparison)	10^{14}	(9000)	2.4	Pure water
Diamondlike films	10^{10}-10^{13}	(5000) and more	2.0-2.8	"
Carbon-containing films	10^9	2000 and less	2.0	Yellowish-brown
Graphite films	10^{-7}	Very low	–	Black

The main characteristics of the sputtering and deposition processes for carbon films with diamondlike properties are given in Table 7.1. There are several metastable phases of carbon: diamond, graphite, cubic diamond, and α- and β-carbine. The energy of the deposited particles has the greatest influence on the properties of carbon films. The optimum particle energy is in the range 10-50 eV.

Deposition of fast ions from carbon plasmas. During production of carbon films by deposition of fluxes of carbon plasma with ion energies of 20-1000 eV, it has been found that the density grows linearly from 2.7 to 4 g/cm^3 when the ion energy is raised within the range 20-50 eV [233]. At energies of 50-70 eV the density of films with thicknesses of tens of microns remained constant, and at ion energies above 70 eV it decreased rapidly, reaching saturation at 100-200 eV.

A film density of 4 g/cm^3 exceeds the density of cubic graphite (3.5 g/cm^3) and is close to the density of the hypothetical metallic form of carbon. For ion energies of 0.1-25 eV graphite and carbine predominate in the film; for energies of 25-40 eV, carbine and diamond; and for energies of 40-70 eV, diamond and the C_s phase. When the ion energy exceeds 70 eV, the amount of less-dense phases in the film increases. This is related to the buildup of vacancies. The formation of the dense phases is explained by the accumulation of interstices and the transfer of sufficient energy from the incident ions for local phase transitions.

Deposition of carbon ion beams. Intense ion beams are produced at relatively high accelerating voltages (see Chapters 1-3). In order to reduce the energy of the ions and maintain their intensity, an ion slowing-down (decel) system is used. Chaikovskii et al. [143] have used a duoplasmatron and an ion accel–decel system to produce carbon beams with energies of 20-100 eV and currents of up to 6 mA in a working vacuum of 2.3·10^{-5} Pa at a distance of 700 mm from the ion source. Films of carbon were deposited on Ni and Si monocrystals with [111] orientation. The thickness of the films was 0.1-1.5 μm. In color they ranged from light gray to dark brown, and they had high adhesiveness and were slightly scratched by sapphire. Their specific resistivity varied over the range 10^7-10^{10} $\Omega \cdot cm$. In order to make the diamondlike films clear another deposition method was used which was based on sputtering graphite with a single beam which is deposited on a substrate that has been bombarded by another ion beam.

The two-ion-source method has been proposed by Vaismantel et al. for producing clear diamondlike films [143]. The first source, a Finkelshtein-type ion source, produces a beam of argon ions with energies of up to 10 keV and a current density of 0.5-1.0 mA/cm² at a pressure of $7 \cdot 10^{-5}$ Pa which sputters a carbon target. The rate of film growth is 0.5-1.0 nm/min for a target-to-substrate distance of 15 cm. The films are made clear by a beam of argon ions or a mixture of argon and methane from the second source with an accelerating voltage of up to 2 kV and an ion current of 0.2-5 mA. The rate of growth of the film thickness (including sputtering) is 0.5-1.0 nm/min. In order to increase the deposition rate and adhesion of diamondlike films, benzene vapor at a relatively high pressure was used.

Plasma-chemical deposition of films from benzene at a working pressure of 10^{-1} Pa and with an accelerating voltage at the substrate of 0.1-5 kV has yielded deposition rates of over 60 nm/min [143]. Films with thicknesses of several microns and good adhesion were obtained on glass, steels, Si, and NaCl.

Some electrical and physical properties of carbon films are listed in Table 7.2.

A single ion beam can be used to perform the operations of sputter cleaning the substrate, sputter deposition of a carbon coating, and improving the physical properties of the coating by ion bombardment.

A method for sputter deposition of materials with simultaneous modification of the properties by ion bombardment. Ion-beam sputtering has been used to make diamondlike films at the NASA Lewis Research Center. An argon ion source with an 8-cm-diameter argon ion beam was operated in a vacuum vessel with a diameter of 1.5 m and length of 4 m [235]. The ion current was 55 mA for an energy of 1000 eV. A large 30.5×30.5 cm² pyrolytic graphite target was sputtered. Carbon films were deposited on copper, tantalum, and fused quartz surfaces after they were sputter cleaned by ion bombardment or by simultaneous sputtering and bombardment with argon ions. The samples were rotated at 5 rps in order to ensure uniform deposition. The ion beam was directed onto the deposition sample through a graphite comb whose transparency could be varied from 0% to 50%. The samples and the comb were positioned along the axis. Ion cleaning lasted 5-300 s. For a comb transparency of 50%, 0.7 carbon atom on the average were sputtered per deposited carbon atom. The carbon ion current density varied over the beam radius from 0 to 2 mA/cm². The average rate of sputtering of the fused quartz during the surface cleaning period was 0.14 nm/s, and the average deposition rate was 0.0275 nm/s for a comb transparency of 0, 0.0213 nm/s for 25%, and 0.00954 nm/s for 50%.

The sputter-deposited carbon films had a density of 2.1 g/cm³, and with additional ion bombardment through a 25% transparent comb, 2.2 g/cm³. Pure monocrystalline graphite has a density of 2.26 g/cm³. Films of silicon, chromium dioxide, and silicon carbide had an amorphous structure, and their density varied from 2.2 to 3.2 g/cm³. The specific resistance of the carbon films deposited by sputtering and ion bombardment (comb transparency 25%) exceeded 10^{11} Ω·cm.

The optical absorption, reflection, and transmission of the films obtained by both methods were very similar. For example, for 170-nm-thick films both processes yielded reflectivities of 0.2, absorptions of 0.7, and transmissions of 0.1 at a wavelength of 555 nm.

Production of diamond by implantation of carbon ions in natural diamond. In 1980-1983 a method of obtaining diamonds by implanting 10-40 keV C^+ ions in a heated crystal of natural diamond was investigated [236]. Unlike the surface layer method [143], this method is based on the implantation of ions at a depth of hundreds of atomic layers below the surface of a crystal, as well as on radiation damage that leads to growth of the diamond.

Sputtering of graphite and diamond by carbon ions is relatively insignificant, since for energies of 1-50 keV the sputtering yield is roughly 0.1 atom/ion. Radiation damage is extremely important since 10-keV C^+ ions create a defect concentration considerably greater than the concentration of implanted atoms (by three orders of magnitude at the surface of diamond and by a factor of 20 at a depth of 15 nm). The density of dislocations exceeded 10^{15} mm^{-3}. However, at certain irradiation temperatures and dose accumulation rates diamond layers with thicknesses of 15-45 μm can be grown.

A 5×5 mm^2 piece of pure polished diamond was irradiated through a silicon mask that served as a demarcation line, at temperatures of 300-800°C by 100-keV C^+ ions at dose accumulation rates of 10^{17}-10^{18} m$^{-2} \cdot$s^{-1}. At low deposition rates (0.01-0.1 μm/h) diamond layers of thickness 0.1-0.3 μm were formed at temperatures of 350-700°C. At 300°C soft amorphous layers were obtained and at 800°C, hard amorphous layers. For fluxes of 30-keV ions above $5 \cdot 10^{19}$ m$^{-2} \cdot$s^{-1} and carbon deposition rates of 0.5-5 μm/h, diamond layers were obtained at temperatures of 620-1050°C and their thickness was 0.8-15 μm. Under these conditions diamond formation did not depend on the energy of the carbon ions over 10-40 keV or on the angle of incidence over 0-85°. Diamond growth was observed on the (111), (100), or (110) planes of cleaned and polished natural surfaces of chiastolite.

The physical and chemical properties of synthetic diamond are essentially identical to those of natural diamond. The mechanism for diamond synthesis consists of sequential accumulation and migration of point defects in dislocation structures.

Diamondlike films have found application in the manufacture of thin film transistors, for raising the wear resistance of cutting instruments, and for making protective coatings on laser diodes. These films will evidently find extensive application in instrument building, electronics, optics, and machine building [143]. Diamondlike films obtained by ion deposition at pressures of 10^{-3}-10^{-4} Pa on CR39 plastic lenses raised their scratch resistance by a factor of 3-6. Similar deposits on industrial cutting blades has made it possible to reduce the force needed to cut teletype paper moving at 2.0 m/min by a factor of 2.5-4 and to increased by 10-100 times the operating lifetime over which a blade loses its cutting properties [238].

7.3 Production of Films of Metals, Alloys, Oxides, Borides, Carbides, Nitrides, Silicides, Sulfides, Selenides, and Tellurides by Cathode Sputtering

We now consider some examples of the application of cathode sputtering for producing thin films of metals, alloys, and chemical compounds in equipment manufactured by Leybold-Heraeus GMBH of West Germany [239].

Diode cathode sputtering systems have two disadvantages: (1) a low deposition rate, only about 10% of that by vapor deposition, and (2) samples mounted on the anode are subject to electron bombardment and are heated to 300-500°C. For a discharge voltage of 1-5 kV the power density at the cathode is 3-10 W/cm^2.

When a magnetron (or high-speed) sputtering system is used, the sputtering rate is 5-10 times higher and the substrate temperature is reduced to 50°C. The discharge voltage is reduced to 300 V, but the thermal load on the cathode increases to 30 W/cm^2. Using magnetic fields with different configurations makes it possible to deposit films on surfaces of different shapes: plane, ribbon, cylindrical.

Typical deposition rates when argon is used are 1 μm/min for aluminum, 2.5 μm/min for copper, and 0.1 μm/min for quartz, while in an ordinary diode system the sputtering rates are 0.12 μm/min for copper and 0.025 μm/min for quartz. The deposition rates for metals and quartz are proportional to the power or energy flux at the cathode (Fig. 7.1).

The sputtering parameters for deposition of films of different metals on the PK500L system are shown in Table 7.3. The overall width of the coatings obtained on this system is up to 40 cm, and within 32 cm the thickness variation of the films was ±2% without the use of correcting screens. The nonuniformity of the film over the full width is within ±8%. High-speed sputtering sources with powers of 50 and 120 kW have been developed.

Leybold-Heraeus makes sputtering cathodes with diameters of 28, 75, 150, and 200 mm and thicknesses of 1.5-6 mm. The target materials include the following:

Metals: Al, Sb, Cd, Cr, Cu, Au, Hf, In, Mn, Mo, Ni, Nb, Pd, Pt, Se, Ag, Ta, Te, Ti, V, W, Sn, Zn, Zr.

Semiconductors: Si, Ge.

Oxides: Al_2O_3, Sb_2O_3, $BaTiO_3$, Bi_2O_3, Bi_2TiO_5, $Bi_4Ti_3O_{12}$, CeO_2, Cr_2O_3, HfO_2, In_2O_3, La_2O_3, MgO, MoO_3, Nb_2O_5, SiO_2, SiO, $SrTiO_3$, $SrZrO_3$, TaO_2, ThO_2, TiO_2, SnO_2, WO_3, ZnO, Y_2O_3, ZrO_2.

Borides: CrB_2, HfB_2, LaB_6, Mo_2B_5, NbB_2, TaB_2, TiB_2, VB_2, W_2B_2, ZrB_2.

Carbides: B_4C, Cr_3C_2, HfC, Mo_2C, NbC, SiC, TaC, TiC, VC, WC, ZrC.

Fluorides: AlF_3, BaF_2, CaF_2, CeF_3, LaF_3, LiF, MgF_2, Na_3AlF_6, ThF_4, YF_3.

Nitrides: AlN, BN, HfN, NbN, Si_3N_4, TaN, TiN, VN, ZrN.

Silicides: $CrSi_2$, $HfSi_2$, $MoSi_2$, $NbSi_2$, $TaSi_2$, $TiSi_2$, VSi_2, $ZrSi_2$.

Alloys: CrSiO, Ni–Cr: 80/20, 70/30, 60/40, 50/50.

Permalloy: 79% Ni, 16.7% Fe, 4% Mo, 0.3% Mn; W–Ti 90/10; In–Sn 95/5, 90/10, 80/20.

Sulfides, selenides, and tellurides: As_2S_3, CdSe, CdS, CdTe, PbSe, PbS, PbTe, $MoSe_2$, MoS_2, $MoTe_2$, $NbSe_2$, NbS, $NbTe_2$, $TaSe_2$, TaS_2, $TaTe_2$, WSe_2, WS_2, ZnSe, ZnS, ZnTe.

The purity of all these materials is 99.5-99.999.

Applications include:

Electronics: semiconductor films; films for soldering; transparent and electrically conducting films; resistor-contacts, ion-plated films; metallizing and passivating films for high-level integrated circuits and semiconductor components; RCL (resistive-capacitive logic) circuits; quartz resonators.

Optics: interference filters; heat-reflecting mirrors; laser mirrors; beam splitters; reflecting and antireflecting coatings.

Special purpose films: solar protection coatings; protective coatings of hard glass on plastic; wear- and corrosion-resistant coatings; decorative coatings; antistatic (nonelectrifying) coatings; insulating coatings.

It should be noted that the materials that can be sputtered include the carbides of refractory metals with extremely high melting temperatures: ZrC (3530°C), TaC (3880°C), and HfC (3900°C).

Laboratory and industrial systems. Among the laboratory systems sold by Leybold-Heraeus we note the *universal laboratory system with a gate valve* (model Z400), the cathode sputtering system Z550, and the Univeks 450 system. The last system matches the highest standards of vacuum technology. A hydrocarbon-free vacuum is obtained by means of turbomolecular and cryogenic pumps. The dimensions of this system are (lwh) $865 \times 800 \times 1810$ mm^3.

A high-vacuum system for preparation of prescription spectacle glasses, VARICOAT 430, can be used to antireflection coat both sides of the glass without letting air in between runs and to apply antisolar shading and filter coatings as well. In a single load it is possible to coat 14 lenses with diameters of 60 mm or 11 lenses with diameters of 75 mm. The time for antireflection coating the lenses in a single load is about 22 min. Deposition of color films requires about 30 min. In this system the evaporator is an electrical resistance heater.

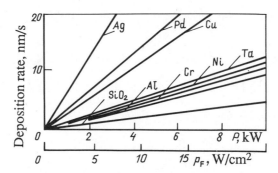

Fig. 7.1. The dependence of the deposition rate for metals and quartz on the power of a magnetron discharge and the power density at the cathode.

TABLE 7.3. Rates of Sputtering and Film Thickness Growth of Metals in the PK500L System for a Discharge Power of 10 kW and Substrate Displacement Speed of 10 cm/min [239]

Material	Sputtering yield, atoms/ Ar ion*	Steady state deposition rate, μm/min	Increase in film thickness in one pass, μm
Ag	3.4	3.90	4.29
Al	1.2	0.66	0.73
Au	2.8	3.0	3.3*2
Cr	1.3	0.72	0.79
Cu	2.3	1.68	1.85
Mo	0.9	0.9*2	1.0*2
Nb	0.65	0.7*2	0.8*2
Pd	2.4	2.2*2	2.4*2
Pt	1.8	1.5*2	1.6*2
Rh	1.5	1.3*2	1.4*2
Si	0.5	0.6*2	0.7*2
Ta	0.6	0.69	0.76
Ti	0.6	0.36	0.40
W	0.6	0.6	0.70
Zr	0.75	1.1*2	1.2*2

*From published data for 600-eV Ar^+ ions.
*2Calculated from the sputtering rate.

The single-vessel cathode sputtering system for two-sided deposition, model Z700P, can be used to deposit TiN coatings as a "gold substitute" on watch cases, bracelets, dial faces, spectacle frames, etc. The low substrate temperature makes it possible to coat plastics. Titanium nitride is deposited on high-quality instruments such as cylindrical and end mills, drill bits, taps, etc. The system weighs about 1500 kg.

The cathode sputtering system for coating powdered materials, model Z750S, takes optimal account of the specifications for production of Ni/Cr metal film resistors. It can be used to deposit a second layer, such as SiO_2, by means of rf sputtering. The dimensions of this system are (lwh) $504 \times 300 \times 200$ mm^3.

The system for cathode sputtering with a gate valve, model Z900, is intended for thin film coating on magnetic heads. Rf etching, sputtering, and rf sputtering with a bias voltage are employed in this system. A low substrate temperature is maintained during the entire time coatings are being deposited. Cryogenic pumps and microprocessor control are used. 44 substrates of area 50.8×50.8 mm^2 or 27 of diameter 3 in. (76.2 mm) can be coated in a single load. The system ensures production of high-quality films as well as reliable and efficient operation.

The modular cathode sputtering systems Z550, Z650, and Z600 are constructed from standardized modules which can be used to expand from simple to complex laboratory installations. They employ rf sputtering using up to three targets, a target bias voltage, rf etching, deposition with the aid of thermal or electron-beam evaporators, and magnetron sputtering. The sizes of the receiver chambers vary from an inner diameter of 600 mm with a height of 100 mm up to (width, depth, height) $580 \times 750 \times 158$ mm^3. This equipment is also used for commercial batch production.

The modular production line system with vertical substrate mounting (model ZV1200) with a "continuous substrate flow" is based on rf sputtering, dc sputtering, magnetron sputtering, sputtering with a bias voltage, and heating. Single-sided or two-sided deposition of single or multilayer coatings can be done.

Continuous flow systems for deposition of coatings on plane glasses. Over a period of 20 yr Leybold-Heraeus has developed more than 10 commercial systems for coating glass pieces in sizes from 1×2 to 3.18×6 m^2 with antisolar, heat-insulating, and mirror coatings. The model A1400Z3H can handle 165,000 m^2/yr and the A2800Z5H, 2.2 million m^2/yr. Equipment is available for making coatings of up to 45 m in length.

Equipment for coating irregularly shaped pieces with magnetron sputtering, models A900HZ, A1400HZ, and A1900HZ, have vessels with diameters of 900, 1400, and 1900 mm and maximum deposition surface areas of 6.2, 12.5, and 25 m^2, respectively. These systems are used for depositing the mirror layer on reflectors, anticorrosion protection on bolts and other pieces, for metallizing automobile accessories and body parts, making thick metal coatings on sheet metal stock, production of transparent pieces for metallization of upper and lower surfaces, for decorating cast and stamped pieces by combining lacquer or paint finishes with metallic coatings, for metallizing mass production items, production of double-surfaced mirrors, depositing a mirror layer on light reflectors and colored jewelry, and for hardening surfaces.

More than 200 systems for ribbon deposition have been delivered to various countries throughout the world. The model A500B2K, A650B2K, and A800B2K capacitor systems for aluminum deposition are used for depositing films on ribbons with widths 500, 650, and 800 mm and a maximum coil diameter of 420 mm. Films up to 1.5 µm thick are deposited on one or both sides of all capacitor papers. Zinc, as well as aluminum, films can be deposited.

The systems for packing materials, models A1000B2P, A1350B2P, A1600B2P, and A2000B2P, are intended for depositing coatings of aluminum on films and papers with widths of 1000, 1350, 1600, and 2000 mm for decorative and functional purposes.

Planar magnetron sputtering sources (magratrons) with powers of 5, 10, and 20 kW are manufactured by Balzers (Liechtenstein) [240]. The utilization of material is as high as 50%, and the power density at the cathode is 50 W/cm^2 for voltages of 200-900 V. The working pressure range is $5 \cdot 10^{-2}$-5 Pa. The discharge current is 10, 20, or 40 A, and the discharge current is regulated to within better than ±1%. The cathode dimensions are 127×254 and 127×444.5 mm^2.

In the LLS-801 sputtering system, which has a gate valve for loading and is used for metallizing integrated circuits with aluminum, platinum, titanium, and tungsten–aluminum alloy, there are five working positions for mounting planar magnetrons and accessories. This model is equipped with an automatic control system [240a].

In the USA ion sputtering is used by 16 companies, of which 12 use this process for making thin films. 13 British and 17 Australian companies use sputtering processes for manufacturing and utilizing thin films. Besides the above-mentioned two, these companies include Edwards, High Vacuum Equipment Corp., Thin Film Technology, Ulvac, Varian, Riber, and others [190].

Of the Japanese companies, NEVA produces an *automatic system for sputtering*, series ASP-600, and a *system and production line for sputtering*, series ILS-700, ILC-703, and ILC-704 [241, 241a]. In the last series, planar magnetrons with cathode areas of 5×15 in.^{2} (127×381 mm^2) are used. Two dc power supply circuits provide 12 kW of power for sputtering. During a cycle lasting 130 min, 20 substrate carriers are loaded into the machine and covered with 1 μm of aluminum. The effective area of a single carrier is 280×380 mm^2, and the combined area is 21,280 cm^2.

These systems are used in semiconductor electronics for the fabrication of electrically conducting films for electrodes, insulating films, indicator elements in liquid crystals, resistors, solar cells, heads for magnetic disks, etc. In the optical industry they are used for making lenses with antireal processing, mirrors, nonuniformly reflecting mirrors, filters, etc. They are also used for coloring and surface hardening of decorative pieces.

In East Germany the VEB Elektromat in Dresden makes the model HZM-4 *rf system for cathode sputtering* [242]. It is equipped with three coating deposition stations and four holders for blanks on which it is possible to sequentially deposit layers of three different materials, including dielectrics, in a single vacuum cycle. Ion etching and heating of the substrates ensures that the films are firmly bonded to the blanks. The sputtering targets and substrates have diameters of 200 mm. The nonuniformity in the film thickness out to a diameter of 155 mm is ±5 %. The 4-kW rf generator operates at a frequency of 2.5 MHz.

In the Soviet Union Ivanovskii, Maishev, and co-workers [243, 243a] have developed several types of systems for ion-plasma deposition of films and surface machining. These systems have high productivity and produce high-quality films. For an argon pressure of $5.4\text{-}6.7 \cdot 10^{-2}$ Pa the current density is about 20 mA/cm^2, and the deposition rate reaches about 1 μm/min.

It should be noted that the basis on which magratrons operate is Kesaev and Pashkova's idea of electromagnetically stabilizing the location of the cathode spot with a magnetic field [244]. Priority in inventing the magratron belongs to the Soviet research group of Minaichev et al. [244a].

Some applications of magnetron sputtering systems. Besides the above applications of cathode sputtering for producing various kinds of films, some new and promising approaches for the use of magnetron sputtering should be noted.

Metal coatings for microsphere shell-targets for inertial thermonuclear fusion [163]. Metal coatings on small particles have a wide range of potential applications, including substances for accelerating catalytic reactions, fillers for conducting epoxies, and targets for ion and electron beam and laser-initiated inertial thermonuclear fusion. The proposed targets are spherical glass shells with diameters of 100-300 μm and wall thicknesses of 1-20 μm filled with a deuterium–tritium mixture and coated with layers of low or high atomic number metals (beryllium, gold, platinum, tantalum). The surface smoothness and thickness uniformity in the sputtered layers, which are a few microns thick, must be better than 10 nm.

Meyer, Clocker, et al. [163] have obtained coatings of platinum, beryllium, copper, nickel, and aluminum on glass spheres under certain conditions (temperature and gas pressure) using a magnetron sputtering system.

Corrosion-resistant amorphous metallic alloys. Thakoor et al. [165] have obtained amorphous films of the alloy $(Mo_{60}Ru_{40})_{82}B_{18}$ using magnetron sputtering (the S-gun of Sputter Films, Inc.) and high-purity powders of Mo, Ru, and B at an argon pressure of 0.7 Pa. Compared with pure molybdenum in pure sulfuric acid, the potential dynamic polarization curves showed a reduction in the current density for the alloy by three orders of magnitude for voltages in the range 0.5-1.2 V. For voltages between –0.25 and 0 V this difference is considerably less, and from –0.5 to –0.3 V the difference is in favor of pure molybdenum.

Superconducting layers of NbN *by reactive magnetron sputtering.* The highest superconducting transition temperature, 14.2 K, has been obtained by Bacon [165] in a mixture of 15% N_2 and 85% Ar with a total pressure of 1.06 Pa. The substrate temperature was estimated at below 90°C during deposition. The films consisted of extremely thin, rather uniform grains with diameters of roughly 5 nm. Bubbles, if present, were no larger than than 0.7 nm. The grains were oriented isotropically. The parameter of the face-centered lattice was 4.46. The highest published transition temperature for solid face-centered cubic NbN is 4.39 K. Probably these films will be used to make Josephson semiconductor junctions. The problem of obtaining superconducting films by cathode sputtering is the subject of many papers.

Transparent, conducting, and heat-reflecting films on glass and plastics.
Films of In_2O_3:Sn and ZnO have been obtained by Brett et al. [165] by sputtering
15-cm-diameter targets of pure (99.99%) Zn or 90% In + 10% Sn in a magnetron
with an Ar/O_2 atmosphere and deposition on glass or polyester. The sputtering rate
was 75 nm/min. They controlled the sputtering process with a microcomputer, so
they could obtain films with a transmission spectrum similar to the solar spectrum
and a high reflectivity for heat (infrared) rays at room temperature. The thickness
of the first films on Corning 7059 glass was 350 nm. Similar films were also obtained on a 13-μm-thick polyester ribbon. The specific resistance of the best films
was $9 \cdot 10^{-4}$ Ω·cm. The optical transmission of the pure polyester and of polyester
with a sputtered film of ZnO exceeded 90% over the wavelength range 400-800
nm. The conditions for obtaining these films were: power to the cathode 150 W,
substrate potential 65 V, overall pressure 2.0 Pa, and argon and oxygen feed rates
of 2.0 and 1.5 cm^3/min, respectively. When the pressure was reduced to 0.5 Pa,
the transparency of the films fell to 80%.

Coatings in gas turbines for reduction of high-temperature corrosion. The
battle with high-temperature corrosion requires a combination of special chemical
and mechanical properties and materials. Cathode sputtering in diode glow, magnetron, and thermionic discharges is used to obtain metal–ceramic and ceramic–
metal layers. By varying the type of bombarding ion, substrate temperature, sputtering rate, mechanical and other preparation of the substrate, and other parameters,
it is possible to obtain the required grain size and orientation [245].

Binding of radioactive gases. A large sputtering system has been designed to
provide a high sputtering rate. At a power of 50 kW it trapped ^{85}Kr at a rate of 2
nl/h in a film whose composition was approximately $Fe_{0.79}Y_{0.12}Kr_{0.09}$ [245].
The gas absorption was a factor of 2 greater than in earlier devices.

Films obtained by cathode sputtering are used to make solar cells and photo-
chemical converters for photoelectrolysis of water to produce hydrogen, in order to
reduce light scattering and the refractive index of metal oxide layers, to make multilayer interference mirrors and filters, and for many other purposes. Ion sputtering
has great promise in powder metallurgy for removal of contaminants and oxides
from grain surfaces and for the production of alloys [160].

7.4 Ion Sources for Thin Film Sputter Deposition

High-intensity ion sources. In addition to their obvious advantages,
magratrons have some limitations: the discharge and deposition take place at relatively high pressures (on the order of 1 Pa), which causes contamination of films;
there is no possibility of controlling the ion current to the sputtering target without
changing the discharge parameters; it is difficult to reproduce the discharge conditions on changing from one sputtering material to another because of changes in the
secondary emission of ions and electrons; it is difficult to sputter magnetic materials such as iron, nickel, cobalt, etc.; and the utilization efficiency of the target material is relatively low (25-60%).

Ion sources that create intense beams of inert gas ions have made it possible to eliminate all these difficulties. The first high-intensity hydrogen ion source, the duoplasmatron, was developed in the Soviet Union in the early 1950's [246a-c]. In 1961 a duoplasmatron produced a beam of 60-keV protons with a current of 30 mA and a current density of 3 A/cm^2 at the emission aperture, and in 1963 a steady-state beam of hydrogen ions was obtained [246] with an extraction voltage of 65 kV and a current of 850 mA at a target located 2.1 m from the ion source. A single magnetic lens and diaphragms made it possible to isolate beams of atomic or molecular ions containing over 95% of the desired component. At a pressure of 2.7·10^{-3} Pa the space charge neutralization was 99%. The species mix of the H_1^+ and H_2^+ beams changed when the gas feed rate was varied from 0.2 to 0.65 cm^3/s. Under normal conditions the corresponding change in H_1^+ was from 30 to 50% and in H_2^+, from 60 to 46%. The gas utilization efficiency was 30%. In forming a 500-mA cylindrical beam at a distance of 2 m from the source the average ion losses on the electrodes and diaphragms were 6%. Ions were collected from a plasma surface area of 100 cm^2, the electron temperature was on the order of 10 eV, and the density was 1.5·10^{10} cm^{-3}.

A duoplasmatron with a developed (extended) plasma surface has been used to obtain a 210-mA, 70-keV He$^+$ ion beam, a 420-mA, 80-keV hydrogen ion beam (calorimetric measurements), and a 35-mA, 60-keV argon ion beam (electrical measurements) [247].

The shortcoming of these sources lay in their comparatively large size and mass, as in the case of an arc discharge source located in a strong magnetic field that produced 100-keV hydrogen ion beams with steady-state currents of up to 0.75 A and pulsed beams with currents of about 2 A [248]. The source of Lamb and Lofgren weighs about 350 kg and that of Poroshin and Kutan, 140 kg.

A reflex discharge source with a mass of 45 kg yielded hydrogen ion beams with a current of 0.5 A and an energy of 115 keV, as well as 0.15-A, 75-keV helium ion beams [249]. The half angle divergence of a 42-mm-diameter He$^+$ ion beam at a distance of 1.6 m from the source was 1°. The surface energy flux at the crossover was as high as 8 kW/cm^2. The operating pressure of hydrogen and residual gases at the current collector was 2·10^{-3} Pa. The ion emission current density reached 4 A/cm^2 for an accelerating voltage of 8 kV.

A compact 60-mm-diameter, 70-mm-high duoplasmatron with a permanent magnet made of barium ferrite, a 0.8-mm-diameter emission aperture, an arc current of 1 A, and an accelerating voltage of 15 kV produced a hydrogen ion beam with a current of 30 mA and a current density at the emission aperture of 6 A/cm^2 [249a].

The characteristics and principle of operation of the ion sources developed by Kaufman and co-workers have been described previously (see Section 5.3). Further improvements in this type of source have involved enhancements in the ion production efficiency and current density of the ion beam [171]. In the standard source the "energy cost" per extracted ion (eV/ion) is given by

TABLE 7.4. Parameters of Ar$^+$ Ion Beams Obtained with Different Ion-Optical System Electrode Configurations in Kaufman Sources [171]. d_e is the Diameter of the Apertures in the Extraction Electrode, l_{e-a} is the Distance between the Extraction and Accelerating Electrodes, and T is the Transparency of the Grid

Configuration of ion-optical system electrodes	Current density in each aperture (μA/cm^2)	No. of apertures in an electrode per cm^2	"Pure" accelerating voltage V_n (V)	Overall accelerating voltage V_t (V)	$R = \dfrac{V_n}{V_t}$	Average beam current density (mA/cm^2)
Two grids with d_e = 2 mm, l_{e-a} = 1 mm	100	18.5	350	500	0.7	1.85*
Two grids with d_e = 1 mm, l_{e-a} = 0.5 mm	75	74	350–400	500	0.7–0.8	3.1–5.5
Three grids with d_e = 2 mm, l_{e-a} = 0.5 mm	60	18.5	150–400	500	0.3–0.8	1.1*
Three grids with d_e = 2 mm, l_{e-a} = 0.5 mm	60	18.5	600	2000	0.3	8.9*
Single grid with d_e = 0.2-0.2 mm^2, T = 82%	(15.5)	(1600)	20	25	0.8	1.0

*The half angle divergence of the beams was 13°.

$$\epsilon \sim 50\, A_p/A_{ex}, \qquad (7.1)$$

where A_p is the outer surface area of the plasma region occupied by primary electrons and A_{ex} is the projected area of the open surface of the screen (emitter) grid.

The minimum gas pressure or N_0 (number of neutrals per m^3) in the discharge vessel required for efficient source operation is given by

$$N_0 \sim 1.4 \cdot 10^{17}\, A_p/\Omega_p \qquad (7.2)$$

or

$$P \sim 8 \cdot 10^{-4}\, A_p/\Omega_p, \qquad (7.3)$$

where Ω_p is the volume occupied by the primary electrons. These formulas are correct to within a factor of 2. For the typical value $\Omega_p/A_p = 0.01$ m, Eq. (7.3) gives $8 \cdot 10^{-2}$ Pa, in agreement with experimental data.

As noted in Chapter 3, the use of multipole magnetic fields created by permanent magnets located on the inner surface of the gas discharge vessel prevents direct incidence of primary electrons on the anodes and makes it possible to create emission surfaces of different configurations. For example, discharge vessels with multipole magnets have been developed with a diameter of 30 cm and a rectangular cross section of 5 × 40 cm^2. Equation (7.1) predicts an energy expenditure of 300 eV/ion for the first design, in accordance with experimental data for argon (200-400 eV/ion). A radial magnetic field made it possible to build a discharge vessel with a diameter of 3.8 cm and obtain a 3-cm-diameter argon ion beam with enhanced uniformity. The maximum diameter of the source, including insulators and the ion-optical system, is 6.4 cm.

In order to raise the ion current density at relatively low ion energies, studies have been made of two-grid ion-optical systems with standard sized beamlet apertures (2 mm) and small apertures in the ion-optical system electrodes (1 mm), and of three-grid and single-grid ion-optical systems (see Table 7.4).

In a single-grid ion-optical system there is no extraction electrode. The ions are collected and accelerated by one electrode. In a nickel grid with a transparency of 82% square apertures with 0.2-mm sides were made for a total of 100 apertures over 25.4 mm. This ion-optical system was first used to simulate an ionospheric plasma and then for ion-beam reactive etching and oxidation. At an energy of 20 eV the argon ion current density was up to 1 mA/cm^2.

The latest advances include choosing silicon monocrystals for the ion-optical system electrode material and machining high-precision apertures in the silicon.

A method of obtaining intense flows and beams of ions using a magnetized plasma emitter fed into a vacuum from a hot-cathode arc discharge was proposed in 1966 [250]. This method made it possible to obtain hydrogen ion currents of 1-3 A with accelerating electrode voltages of 1-6 kV by bringing the electrode close (1-10 mm) to the boundary of the magnetized plasma. Ion beams with currents of 0.6-0.8 A and energies of 30-40 keV were obtained with different distributions of current density over the cross section and length of the beam and a high equivalent electron perveance, $5 \cdot 10^{-6}$ AV$^{-3/2}$ [250a].

In 1962 Danilov [251] developed a theory for ion space-charge compensation in order to raise the current density above the limit set by Langmuir's law. Based on the two papers cited above, in 1977 a new type of ion source with compensation of the positive space charge inside the accelerating gap [243] and a system for ion-plasma sputtering [243a,b] were developed. This ion source produced argon ion beams with a current of 0.2 A at an energy of 1 keV and 0.5 A at an energy of 5 keV. The pressure in the vacuum vessel was $4 \cdot 10^{-2}$-$1.3 \cdot 10^{-3}$ Pa.

Ion beams for sputtering. The first publication on the direct application of ion beams for sputtering appeared in 1954 [171]. A few papers were published on this topic before 1970. The total number of publications over 1954-1975 was 215, mostly after 1970, and three years later the total had reached 445 [194]. Of these papers, 15 were reviews devoted to the applications of ion beams in science and technology. Good university textbooks on the engineering applications and physical and chemical foundations of electron and ion technology have been written [105, 252].

Intense, broad ion beams can be used to sputter any hard material at a rate of several tenths of a nanometer per second, which is comparable to the rate of rf sputtering. Thin films can be deposited on surfaces with areas of many square

TABLE 7.5. The Specific Electrical Resistance of Films ($\mu\Omega\cdot$cm) Obtained by Different Methods [254]

Film	Ion-beam sputtering	Rf sputtering	Evaporation
Au (300 nm)	5.9	3.1	3.7
Al (400 nm)	28.5	6.5	4.0
Ta (400 nm)	225.8	218.1	233.7
Ni Fe (400 nm)	27.5	17.0	18.0

TABLE 7.6. Stresses in Thin Films (N/m^2) [254]

Film	Ion-beam sputtering	Rf sputtering	Evaporation
Au	$1\cdot10^7$	$5\cdot10^7$	$4\cdot10^7$
Si	$6\cdot10^8$	$2\cdot10^8$	$4\cdot10^8*$
Al	$3\cdot10^7$	$6\cdot10^7$	$2\cdot10^7$
Ni Fe	$8\cdot10^8$	$3\cdot10^8$	$3\cdot10^8*$

Note: The values marked with * correspond to film thicknesses of 100 nm, the other to film thicknesses of 400 nm.

centimeters. There is particular interest in the sputtering of powders, composite materials, and dielectrics. Films obtained by ion-beam sputtering are purer than those produced by rf sputtering. The amount of Ar or Xe in NbTi films lies between 0.03-0.9%. The concentration of oxygen compounds with chemically active metals is reduced to below 1%.

The composition of films obtained by sputtering of multicomponent targets is the same as (or differs little from) the composition of the sputtering target, e.g., NbTi. At the same time, it is possible to selectively or preferentially sputter the components, as of $CdCo_5$. There is a difference in the angular distributions of the sputtered atoms (NiFe), and variations occur in the implantation in the film of bombarding ions reflected from the target. The texture of the sputtering target may greatly change the angular distribution of the sputtered atoms, as is observed in the form of Wehner spots during the sputtering of different cleavage planes of monocrystals [62].

Ag/SiO$_2$ cermets have been obtained by ion-beam sputtering of two targets made of Ag and SiO$_2$. By using sputtering to control the composition of the film, it was possible to change the specific resistance by seven orders of magnitude. Ion-beam sputtering has been used to obtain superconducting films of Nb–Ge, NbTi, and Nb$_3$Al, semiconducting films of Si, GaAs, InSb, and PbSnTe, polymer films, monocrystalline films of PbSnTe on CaF$_2$ and BaF$_2$, and epitaxial films of Si on spinels.

TABLE 7.7. Magnetic Properties of NiFe Films with Thicknesses of 300–400 nm [254]

Film characteristic	Ion-beam sputtering	Rf sputtering	S-gun
Mass composition of film % Ni–Fe	81–19	81–19	81–19
Mass concentration of Ar, %	$\leqslant 0.005$	$\leqslant 0.03$	$\leqslant 2$
Thickness uniformity, %	± 5	± 2	± 5
H_c, A/m	32–40	32–40	40–240
H_K, A/m	~240	400	400
ρ, $\mu\Omega\cdot$cm	25–30	16–18	30–40
$\Delta\rho/\rho_0$, %	1.5–2	3–3.5	1–2
Mechanical stress, N/m^2	$8\cdot 10^8$	$3\cdot 10^8$	–

TABLE 7.8. The Effect of Residual Oxygen on the Sputtering Rate of Various Materials

Material	O_2, %	U_s, nm/min	Material	O_2, %	U_s, nm/min
Au	0	56.0	W	10	8.3
Au	5	59.0	W	15	4.0
Au	10	57.0	Ni	0	14.0
Au	15	59.0	Ni	15	7.0
Pd	0	30.0	Ti	0	3.0
Pd	5	31.0	Ti	15	2.0
Pd	10	32.0	Pt	0	24.0
Pd	15	31.0	Pt	15	10.0
W	0	11.0	SiO_2	0	7.5
W	5	8.8			

Two ion-beam technique. As has already been noted, the two ion-beam technique, in which one beam sputters a target (e.g., two-component A_xB_{1-x}) and the other bombards the film (A_yB_{1-y}) as it grows on a substrate, opens new possibilities. The composition of the film can be varied over wide limits. This is related to the fact that the ratio of the sputtering yields for elementary Cd and Co is 0.37, while the ratio varies from 2 to 6 in their alloys. This change is caused by a difference in the surface binding energy of the two elements and by the implantation of ions in the film. Thus, the implantation coefficient for argon ions in amorphous Cd–Co films varies from 14 to 100% at low irradiation doses.

Simultaneous ion-beam bombardment and deposition have been used to fabricate polished copper films on optical surfaces.

Reactive ion-beam deposition. A single source with a chemically active gas (oxygen, nitrogen) has been used to obtain films of the oxides of Zr, Ti, Ta, Si, and Pb and the nitrides of Si, Al, and Ti. By varying the oxygen pressure in the vessel in front of the target it is possible to obtain films of Pb, PbO, and Pb_3O_4.

7.5. Comparison of Films Obtained by
Sputtering and Other Deposition Techniques

A comparison of contamination in gold and tantalum films with thicknesses of 1-3 μm and masses of 2-10 mg obtained by evaporation and ion-beam sputtering [253] showed that the residual gas pressures during evaporation by the heater and by the electron beam were $6.7 \cdot 10^{-4}$ and $6.7 \cdot 10^{-5}$ Pa, respectively, and $2.7 \cdot 10^{-2}$ Pa during evaporation by sputtering. The average energy of the ions bombarding the target was about 1200 eV, and the ion current density was 100 μA/cm^2. In a massive sample of gold the total amount of metallic impurities (Ag, Cu, Si, Pb, Fe, Mg) was 32 ppm. During evaporation with a heater the amount of metallic impurities in the film rose to 1550 ppm and the amount of gaseous, to 37 ppm. Using a stainless steel cathode and argon raised the amount of metallic impurities to 12,660 ppm. Replacing argon by xenon reduced these impurities to 7170 ppm, and replacing argon by xenon and the stainless steel by tantalum reduced the metallic impurities to 10 ppm (Ag and Cu). The gaseous impurities in the film were Xe (10 ppm) and C (3 ppm). Sputtering of tantalum under these conditions caused the appearance of metallic (6120 ppm) and gaseous (192 ppm) impurities in the films. During electron beam sputtering the films were found to have 10,100 ppm metallic and 130 ppm gaseous impurities (compared with 1350 and 160 ppm in films obtained by ion sputtering).

The deposition rate for tantalum films with a 6-keV, 1-mA xenon ion beam was 12.5 nm/min and for gold films, 30 nm/min. Deposition rates of 20 and 50 nm/min have been obtained during sputtering of manganese and bismuth by argon ions. The IMMI-3 ion micromachining system has been also used to produce films of copper, silver, vanadium, Permalloy (FeNi), and the alloys $ZrZn_2$ and Sc_3In. The target temperatures can be varied over 80-450 K. Films of Au, Ta, and several other metals, compounds, and alloys had stronger adhesion than those obtained by rf sputtering with a constant discharge voltage.

The physical and mechanical properties of metal and alloy films obtained by ion-beam sputtering, rf sputtering, and evaporation are compared in Tables 7.5-7.7.

The ion energy ranged from 300 to 1500 eV, the ion current density was 0.5 mA/cm^2, and the argon pressure was $1.1 \cdot 10^{-2}$ Pa. A Kaufman ion source with a beam diameter of 7.5 cm was used for sputtering. The deposition rates for Al and W had peaks of about 7.5 and 3.0 nm/min at an angle of incidence of the ions on the target of 60°. At 45° the deposition rates with 1000-eV argon ions were roughly 5.0, 7.5, 15.0, and 20.0 nm/min, respectively, for W, Al, Si, and Au.

Peculiarities of film deposition and ion-beam sputtering of surfaces. During thin film deposition and ion milling of microscopic structures, it is necessary to take a number of circumstances and effects into account. These include: scattering of ions along the beam path with subsequent sputtering of the deposited material; the contrast in resolution of segments connected by a junction on which the angle of incidence of the ions is different on neighboring segments; multiple deposition on neighboring segments of the surface; radiation damage; and the action of the ion chemical process [255].

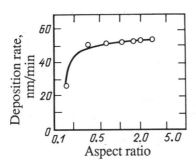

Fig. 7.2. The dependence of the film deposition rate inside through holes in steel on the aspect ratio of the holes for a potential of –500 V on the sputtering target and an argon pressure of 6.7 Pa in the glow discharge [256].

TABLE 7.9. The Sputtering Yield and Rate of Film Deposition on the Inner Walls of Through Holes for Several Materials [256] with an Aspect Ratio of 0.3

Material	Sputtering yield at 500 eV (atoms/ion)	Deposition rate (nm/min)	Material	Sputtering yield at 500 eV (atoms/ion)	Deposition rate (nm/min)
Ag	3.4	65.0	Mo	0.9	13.5
Pd	2.4	50.0	C	0.2	2.0

TABLE 7.10. The Sputtering Yields of a Number of Metals for Sputtering of Ingots by 10-keV Argon Ions at Normal Incidence

Material	Y, atoms/ion	Material	Y, atoms/ion
Co	3.7	Hf	0.4
Cu	8.5	Ta	1.7
Nb	1.4	Pt	4.3
Mo	1.5	Au	4.5
Tb	1.6	Th	0.8
Tm	1.5	U	1.4

The presence of oxygen near a sputtering target causes a reduction in the sputtering rate for most base metals (see Table 7.8). An Ar^+ ion beam with a surface energy flux of 0.155 W/cm^2, a current density of 1-5 mA/cm^2, an energy of about 200 eV, and an angle of incidence of 60° was used for sputtering. For Au and Pd the effect of oxygen is negligible, but for Ni, Pt, W, and Ti the presence of oxygen in amounts of 10-15% leads to a sharp reduction in the sputtering rate (by a factor of 1.5-4.7), apparently owing to the formation of oxide films that are resistant to sputtering. The rates of sputtering of W, Ni, Pd, and Au by pure Ar^+ beams with the above parameters are roughly constant at 11, 185, 32, and 56 nm/min for ions incident at angles of 0-30°(50°). As the angle of incidence is raised above 60° (50°), the sputtering rate decreases rapidly. The sputtering yield as a function of the angle of incidence of the ions had a maximum for gold in the range 30-40°, for tungsten at 45°, for palladium at 45-60°, and for nickel at 60°.

When coating the inner surfaces of through holes in samples mounted on a sputtering substrate, it is necessary to account for the variation in the angular distribution as the ion energy is raised. When the energy of the ions is below 1 keV, the angular distribution of the sputtered atoms lies below a cosine (isotropic) distribution. For energies of 1-10 keV it is roughly a cosine distribution, and for higher ion energies it lies above a cosine distribution [62, 256]. The deposition rate depends on the aspect ratio given by the ratio of the diameter to the height of the hole (Fig. 7.2). For a small aspect ratio (0.15) the deposition rate is 40% of that for aspect ratios of 0.3-3 and also depends on the sputtered material (Table 7.9). These data are of interest for the construction of a porous first wall in a thermonuclear reactor.

Beams of sputtered fast atoms. Beams of gold atoms with energies of 20 eV and fluxes of 10^{13}-$2 \cdot 10^{14}$ cm$^{-2} \cdot$s^{-1} have been obtained [257] by sputtering of monocrystalline targets by 10-keV Ar$^+$ ion beams with current densities of 2 μA/mm^2. The beams of sputtered fast atoms made it possible to obtain thin films of a specified composition with a preferred orientation.

During sputtering of a polycrystalline target the differential sputtering yield varies from 0 to 2.9 atoms/(ion·sr). When the (110) and (100) planes of Au are sputtered, the corresponding maximum values are 1.8 and 2.5 atoms/(ion·sr), while when the (111) plane is sputtered, the directivity of the sputtered atoms increases to 7.5 atoms/(ion·sr), which corresponds to a flux of $2 \cdot 10^{14}$ atoms/cm$^2 \cdot$s.

The temperature at which epitaxial coatings, for example (100), are formed during evaporation in air, vacuum, and superhigh vacuum varies from 150 to 400°C. When ion bombardment is used, this temperature is reduced to 100°C.

7.6. Some Applications of Ion Beams in Thin Film Production

Isotopic targets [258]. Electron beam evaporation coating is used to prepare isotopic targets for high-resolution nuclear spectroscopy. The overall efficiency of evaporation–condensation is 70-80%. The principal defect of this method is contamination of the isotopic target with material from the substrate, which is heated to a high temperature. The contamination of isotopic targets can be reduced by using 10-keV Ar$^+$ ion beams with target and beam dimensions on the order of 1 mm^2. The ion current in a focused beam of this type is 1-2 mA. The pressure in the vacuum vessel is kept at a level of $1.33 \cdot 10^{-4}$-$1.33 \cdot 10^{-3}$ Pa, which ensures that the mean free path of the sputtered atoms is about 5 m. Some measurements taken under these conditions are listed in Table 7.10.

Experiments show that when metal powders are used as targets the angular distribution of the sputtered atoms is much wider than when ingots are used. Therefore, to increase the efficiency of collection of the sputtered isotopes, powders of the latter were placed in a cavity in a graphite block. The cavity was made in a cut of the graphite block that ensured oblique incidence of the ion beam on the surface of the powder target. A cylindrical isotope collector was mounted a distance of 4 cm from the surface and parallel to it. It was possible to obtain nine targets with a standard area of 0.8 cm^2 simultaneously. The thickness nonuniformity varied over 1-5%. The collectors were thin self-supporting films of zirconium, tantalum, and tungsten. Three targets with an average thickness of 60 $\mu g/cm^2$ were obtained from 2 mg of ^{92}Zr placed in the graphite block. Isotopic targets of ^{90}Zr, ^{91}Zr, ^{94}Zr, and ^{96}Zr with thicknesses of 100 $\mu g/cm^2$ were also prepared.

If the collector surface is covered with a thin film of NaCl, then the coated isotopic target can be fished out of water after the salt is dissolved. Several self-supporting isotopic targets with thicknesses of 500 $\mu g/cm^2$ were made from 25 mg of ^{182}W metal powder. Ion-beam sputtering was especially useful for depositing osmium layers on monocrystalline hafnium, which was extremely difficult to do by means of evaporation–condensation.

The efficiency of collecting the sputtered isotopes is as high as 70%. Targets with thicknesses of 100-300 $\mu g/cm^2$ were obtained from 5 mg of isotopically enriched material. Targets of isotopes of Co, Nb, Zr, Y, Hf, W, Os, Pt, Th, and U were made on carbon films with thicknesses of 10-30 $\mu g/cm^2$ for high-resolution spectroscopy. Self-supporting Zr and W foils with thicknesses of 100-500 $\mu g/cm^2$ have been obtained.

Optical waveguides [259]. A 2-keV, 12.5-mA argon ion beam with a diameter of about 7 cm was aimed onto a tantalum target through an aperture in a 50.8-mm-diameter tantalum diaphragm. The sputtered atoms caused getter pumping, and the partial pressures of the residual gases (water vapor and oxygen) were kept at levels of $6 \cdot 10^{-6}$ and $5 \cdot 10^{-8}$ Pa, respectively. The ratio of the number of Ta atoms to the number of H_2O molecules falling on a target of pure Corning 7059 glass was roughly 10, and the pressure of the sputtered gas was about $6 \cdot 10^{-3}$ Pa.

Metal films having a body-centered cubic structure with a predominantly (111) facet orientation were produced using an Ar^+ beam. Even at a low deposition rate (about 2.5 nm/min), the temperature coefficient of resistivity was about 1200 ppm/°C, while for massive tantalum, it was 3960 ppm/°C.

Transparent amorphous films have been obtained with the aid of O^+ and Ar^+ ion beams. For example, when the deposition rate was 1.5 nm/min on a 1-cm^2 substrate, Ta and O atoms were incident at rates of $9 \cdot 10^{13}$ and $2.25 \cdot 10^{14}$ s^{-1}. The reflectivity of Ta_2O_5 films varied over the range $(2.12-2.148) \pm 0.003$. The maximum reflectivity obtained in a glow discharge in an O_2/Ar mixture was 2.085. The values obtained with anode and thermal oxidation were 2.195 and 2.221, respectively.

The lowest optical losses in these optical waveguides were 11 dB/cm, which is considerably greater than the level obtained with a glow discharge in an O_2–Ar mixture, ~1 dB/cm.

Metal–dielectric–semiconductor structures [260]. Insulating layers of SiN_xO_y and AlN_xO_y have been obtained by reactive etching of silicon or aluminum targets with a nitrogen ion beam. The ratio y/x was varied by introducing pure molecular oxygen near the target. The working gas pressure was $1.3 \cdot 10^{-3}$-$1.3 \cdot 10^{-2}$ Pa. A cold cathode source with a saddle-shaped magnetic field generated a 5-keV, 500-µA beam. The temperature of the substrate was kept around 300°C. The specific resistance and breakdown strength of films with thicknesses of about 100 nm increased as the ratio y/x was raised and had typical values of 10^{16} Ω·cm and $2 \cdot 10^6$ V/cm. A dramatic reduction in the density of states in the interface from 10^{14} to less than 10^{11} cm^{-2}·eV^{-1} in AlN_xO_y was obtained by subsequent processing in a hydrogen plasma and annealing in nitrogen at 300°C. This reduction is explained by the formation of weakening and strengthening Si–H bonds.

Ceramic films with ferroelectric and piezoelectric properties [161]. Films with a nominal composition $Pb(Zr_xTi_{1-x})O_3$ used in electro-optical and converter applications have been produced at an initial pressure of $2 \cdot 10^{-4}$ Pa. Afterwards argon was fed into the ion source and oxygen into the vacuum vessel to a partial pressure of $1.33 \cdot 10^{-2}$ Pa. A 2-keV argon ion beam with a current density of 1 mA/cm^2 was used. Depending on the temperature of the NiCr–Au substrate deposited on Pyrex, the dielectric constant varied from 40 to 100-350 at room temperature. The coercive field for the films was 25.2 kV/cm compared to 6.5 kV/cm for massive PZT ceramic.

Magnetic films by ion-beam sputtering. Magnetic metal films are very widely used in microelectronics. Vacuum evaporation, electrolytic precipitation, and cathode sputtering techniques are used to fabricate them [261]. The use of ion-beam sputtering to produce magnetic films permits more independent variation of the ion energy, sputtered particle energies, and angle of incidence than with rf sputtering.

The two ion-beam method has been used to deposit 100- and 1000-nm-thick films of $Co_{82}Cr_{18}$ on glass, titanium, chromium, and amorphous Ta–W–Ni alloy [262]. The first beam cleaned the substrate for 8 min before the film was deposited with an ion current of 100 mA and an energy of 300 eV. The second beam, with a diameter of 6.5 cm, a current of 150 mA, an energy of 1200 eV, and an argon pressure of $2.7 \cdot 10^{-2}$ Pa, sputtered a target containing 82% cobalt and 18% chromium. Films with thicknesses below 1000 nm deposited on glass had a predominant (100) crystal orientation. Films with a (002) orientation were also obtained. The ratio of the reflected intensity from these planes (I_{100}/I_{002}) fell from 15 to 5% when the film thickness was changed from 100 to 600 nm. A (002) texture developed in thin films on substrates of titanium and amorphous Ta–W–Ni alloy.

Chapter 8

ION FILM DEPOSITION AND
PLATING OF SURFACES

8.1. Optimum Energy for Ion Deposition in Films

In 1955 Rukman, Yukhvidin, and Kalyabina [263] first proposed for the purpose of obtaining uniform-thickness films that the material to be deposited initially be ionized and then directed onto the surface to be coated in the form of an ion beam controlled by a deflection system. The energy of the ions would have to be several tens of electron volts. A 1-cm^2 0.5-μm-thick silver film requires an ion dose of 0.5 C and irradiation for about 1.5 h by a 100-μA beam.

Ion beams, rf discharges, glow discharges, and other methods of producing and accelerating ions are used to deposit films of metals and dielectrics.

The thickness of the deposited films and their density and crystalline orientation depend on the ion energy. Films of lead and magnesium have been produced by deposition of 24-300 eV ions obtained by decelerating 4-keV mass-analyzed Pb$^+$ and Mg$^+$ ions. The optimum ion energy for obtaining dense films is in the range 50-75 eV (Table 8.1).

For a Pb$^+$ ion current of 9-12 μA, the irradiation dose was 0.25 C and the deposition time was 5.5-7 h at a pressure of $1.3 \cdot 10^{-6}$ Pa. For an Mg$^+$ ion current of 12-15 μA, the dose was 0.25 C and the sputtering time 4-4.5 h at a pressure of $6.7 \cdot 10^{-6}$ Pa.

8.2. Vapor Condensation Combined with Ion
Bombardment (Ion Plating) of Surfaces

The disadvantage of using ion beams for sputter deposition of thin films is a relatively low rate of sputtering and film deposition, and the main advantage is a high velocity of the sputtered atoms and a high activation energy for deposition and sorption of atoms on the substrate. Evaporation can proceed at a high rate, but evaporated atoms are slow, with velocities tens or hundreds of times lower than those of sputtered atoms and many orders of magnitude lower than those of beam ions. In order to use the positive aspects of the evaporation and sputtering processes, a method has been proposed for using ion bombardment to stimulate the condensation of vapors produced by electron bombardment [171].

Another method employs an electron beam to ionize the material vaporized by the beam itself. Then a negative potential is applied to the substrate to accelerate the ion component of the evaporated material. This method has been employed successfully by Belevskii and co-workers [108]. The fraction of ionized components in the particle flux approaches 40%. This method, known as thermionic deposition, has produced films of Mo, V, Nb, and Ta with densities and specific resistances close to those of solid (massive) samples.

Vapor condensation with ion bombardment stimulation has yielded aluminum films whose specific resistance changed by over 18 orders of magnitude (from 10^{-6} to 10^{12} $\Omega \cdot$cm) during bombardment with 5-keV O_2^+ ions at doses of from 10^{19} to 10^{22} cm^{-2}. Films obtained at doses above 10^{21} cm^{-2} were made up of small droplets of metal in a dielectric medium or of pure dielectric.

Superconducting NbN films are obtained by N_2^+ ion bombardment during evaporation of Nb.

Ion bombardment during evaporative film deposition can enhance the adhesion of the films, create internal tension or compression forces or remove these forces, and modify the structure (ion-beam annealing) through formation of thermal peaks or the hardening of a phase at high pressure (as a result of a supersonic shock wave). Ion bombardment makes it possible to reduce the formation temperature for epitaxial layers to 30-120°C, to obtain nitrides, etc.

A significant advance in research on ion deposition techniques for various metals and their compounds began with the work of Mattox published between 1963 and 1973 [226, 265]. This method was named ion coating or plating. It is applicable to atomic film deposition processes in which the substrate is subjected to an ion flux with a high enough energy to cause significant sputtering prior to and during deposition. In order to obtain good adhesion it is fundamentally important that (1) a pure surface be created and maintained by sputtering until the film is completely formed, (2) an energy flux be created which is high enough to heat the surface without heating the bulk mass of the sample in order to enhance diffusion, chemical reactions, etc., and (3) the sputtered surface and interface be modified so as to create defects and physical mixing of film with the substrate material.

8.3. Alternating Ion–Vapor Film Deposition

A new, highly efficient method for alternating ion–vapor deposition of films and a theoretical calculation of the activation energy for sputtering have been discussed by Schiller et al. [266]. The activation energy is defined as the ratio

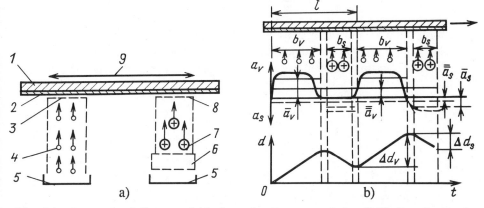

Fig. 8.1. A conceptual diagram of (a) alternating ion–vapor plating [266] (1, substrate; 2, film; 3, region of vapor action; 4, vapor atoms; 5, vapor source; 6, ion source; 7, ions; 8, region of ion action; 9, relative motion of the substrate) and (b) the growth in thickness d of a film during alternating vapor–ion plating (a_v, a_s are the rates; $\bar{a}_v$, $\bar{a}_s$ are the average rates relative to the width b; and $\bar{\bar{a}}_v$, $\bar{\bar{a}}_s$ are the average rates relative to the period length l; v is the velocity of the substrate).

TABLE 8.1. Density and Thickness of Lead and Magnesium Films Made by Depositing Pb^+ and Mg^+ Ions on Graphite and NaCl [264]

Ion	Ion energy, eV	Ratio of film density to solid density, %	Average film thickness (±20 nm), nm
Pb^+	24	64	256
	48	97	190
	72	101	128
	121	80	85
	169	35	50
Mg^+	24	83	190
	48	100	190
	100	67	190
	120	50	140
	170	17	–
	300	15	–

TABLE 8.2. The Density of Chemically Active Centers in 1-μm-Thick Nickel Films on Steel Substrates Obtained by Different Methods at $t = 130°C$ [266]

Film deposition technique	Deposition conditions	Density of chemically active centers, cm^{-2}
Vacuum evaporation	$2.7 \cdot 10^{-3}$ Pa	700 – 800
Deposition in a gas	2.7 Pa Ar	1000 – 1200
Alternating ion plating	2.7 Pa Ar, $a_s/a_v = 0.5$	25 – 30

$$\epsilon = (W_I + W_v)/W_v = (n_I \omega_I + n_v \omega_v)/(n_v \omega_v), \tag{8.1}$$

where W_I and W_v are the energies transferred to the surface by ions and evaporated atoms, respectively; n_I and n_v are the number of ions and evaporated atoms arriving on unit surface per unit time; and ω_I and ω_v are the kinetic energies of the ions and evaporated atoms. Here

$$\omega_v \approx (3/2)k\,T_v \tag{8.2}$$

and

$$\omega_I = e\langle U_I \rangle, \tag{8.3}$$

where k is Boltzmann's constant, T_v is the vaporization temperature, and $\langle U_I \rangle$ is the average accelerating voltage. Usually $W_v \ll W_I$, so that

$$\epsilon \approx 6 \cdot 10^{-3}\, \frac{\langle U_i \rangle}{T_v}\, \frac{n_I}{n_v}. \tag{8.4}$$

During evaporation $\omega_v \approx 0.2$ eV and $T_v \approx 2000$ K. For the sputtered atoms $\omega_s = 1\text{-}2$ eV and for the bombarding ions $\omega_I = 50\text{-}5000$ eV. In the first case $\epsilon = 1$, in the second 5-10, and in the third the energy efficiency may vary from 1 to 2500, depending on the ion energy and the ratio n_I/n_v.

As the energy of the ions bombarding the sputtered substrate is raised, however, reverse sputtering of the condensed substance occurs and the deposition rate is reduced, i.e.,

$$a_{\text{eff}} = a_D - a_s, \tag{8.5}$$

where a_{eff}, a_D, and a_s are the effective mass sputtering rate of the material per unit time, the rate of condensation without ion bombardment, and the rate of sputtering of the film.

The principle of alternating ion–vapor film deposition (plating) involves the action of ion bombardment and vapor of the material at every point of the surface over time (Fig. 8.1a). The vapor and ion sources are separated in space and act on each point of the surface alternately because the substrate is shifted toward these sources. During one displacement period the thickness of the film increases by Δd_v. During bombardment by a beam of ions with sufficient energy the thickness of the film decreases by Δd_s. The thickness of the film increases with time (Fig. 8.1b). The partial growth in the film during a single period depends on the mean deposition rate $\langle a_v \rangle$, the width of the vapor deposition zone b_v, the amplitude l of the displacement of the substrate, and the frequency f of changes in the substrate position:

TABLE 8.3. Film–Substrate Bonds Obtained by Ion Plating [265]

Corrosion-resistant coatings	Adhesive bonds	Electrical contacts
Al, Cr, Au–U	Al, Ag – Be	Au – Be/Cu
Cr, Cd, Al, Au – steel	Cu, Au, Ag – W	Au – TaN, Ta, Cr
Stainless steel	Mo, Ta, Nb, Ni, Au – Ti	Au, Ag – Al, Mo, Nb
Al – Ti	Cu, Sn, Al, Ti, Cr – ferrites	Al, In, In/Ga – CdS
Cr – Mo	Al, Cu – garnet, Pd – stainless steel	Al, Cu, Cr – Si, Ge, PtSi – Si

$$\Delta d_v = f^{-1} (b_v / l) \langle a_v \rangle. \tag{8.6}$$

Thus, the partial sputtering rate of the film is given by

$$\Delta d_s = f^{-1} (b_s / l) \langle a_s \rangle, \tag{8.7}$$

where b_s is the width of the region in which the ions interact with the substrate and $\langle a_s \rangle$ is the average sputtering rate. The film growth over a single period is given by the difference

$$\Delta d = f^{-1} \left(\frac{b_v}{l} \langle a_v \rangle - \frac{b_s}{l} \langle a_s \rangle \right). \tag{8.8}$$

Introducing the average values of the rates over a period, we obtain

$$\Delta d = f^{-1} (\langle\langle a_v \rangle\rangle - \langle\langle a_s \rangle\rangle). \tag{8.9}$$

From this equation we find the frequency of alternating ion plating

$$f = \frac{\langle\langle a_v \rangle\rangle}{\Delta d} \left(1 - \frac{\langle\langle a_s \rangle\rangle}{\langle\langle a_v \rangle\rangle} \right). \tag{8.10}$$

Assuming that $b_s = b_v$, we find that

$$\frac{\langle\langle a_s \rangle\rangle}{\langle\langle a_v \rangle\rangle} = Y n_I / n_v. \tag{8.11}$$

The increase in the film thickness during a single period can be expressed in terms of the number of atomic layers k_1 and the average diameter D_A of an atom as

$$\Delta d = k_1 D_A. \tag{8.12}$$

The necessary condition for alternating ion plating has the form

$$f = (\langle\langle a_v \rangle\rangle / k_1 D_A)(1 - Y n_I / n_v). \tag{8.13}$$

The limiting values of the frequency and the ratio n_I/n_v are

$$f_{max} = \langle\langle a_v \rangle\rangle / D_A \tag{8.14}$$

and

$$10^{-3} \leqslant n_I/n_v \leqslant 10^{-1}. \tag{8.15}$$

These limiting calculations show (for $D_A = 0.1$ nm and a commercially acceptable frequency, say 10 Hz) that the expected deposition rate for alternating ion–vapor plating is 0.3-30 nm/s.

If a disk-shaped holder rotating at 600 rpm is used and a single sandwich layer with a thickness of 1-10 atomic layers is produced between two successive ion irradiation periods at an evaporation rate of 5-50 μm/min and an ion current of 0.1 A/cm², so that $n_I/n_v = 1/10$, then it is possible to obtain deposition rates in commercial equipment of 60-600 nm/min or even 20 μm/min with an activation energy of 100 times that for vapor deposition.

Table 8.2 shows a comparison of nickel films deposited by different methods on steel substrates. The optical reflectivities of nickel films deposited on bronze and steel by alternating ion plating were 38 and 52%, respectively, and were higher than for ion coating in argon at a pressure of 2.7 Pa (19 and 23%, respectively) or for vacuum evaporation (32 and 44%). The corrosion resistance of nickel coatings on bronze produced by these methods, as measured by the change in optical transmission, was the greatest for alternating ion–vapor plating (5%), which is 6-10 times greater than vapor deposition or ion deposition at high Ar pressures [265].

Research and development in ion plating. Ion plating is used to enhance adhesion and to produce corrosion-resistant coatings and electrical contacts (Table 8.3). Ion plating is used to deposit metals, alloys, and compounds on metals, insulators, or organic materials. A bibliography on this topic includes more than 170 papers [171].

Pseudosolutions of inert gases in metals. A large cycle of experimental and theoretical work aimed at producing films with enhanced physical and chemical properties has been carried out by Radzhabov [135].

Simultaneous ion bombardment and deposition has made it possible to obtain high concentrations of implanted particles in solids with binding energies close to the chemical bond strengths; i.e., the possibility of obtaining solid pseudosolutions of inert gases such as titanium argonide, scandium heliumide, etc., has been demonstrated.

The differential equations describing ion implantation, sputtering, diffusion, and capture of impurity gas atoms by lattice defects during ion bombardment of films as they grow on a substrate of the same material have the form [267a, b]

$$\frac{\partial c_f}{\partial t} = D \frac{\partial^2 c_f}{\partial x^2} - v \frac{\partial c_f}{\partial x} + Q(x) - \frac{D}{L^2} c_f, \tag{8.16}$$

where $c_f(x, t)$ is the concentration of free (mobile) atoms, uncaptured by defects, of the implanted impurity at depth x at time t; D is the diffusion coefficient for the implanted impurity atoms; v is the rate of growth of the film, equal to the difference between the rates of deposition and sputtering; $Q(x)$ is the distribution of implanted atoms along their paths; and L is the mean distance between defect-traps.

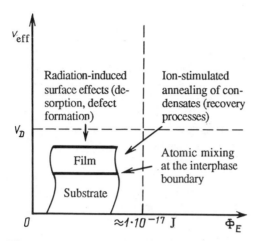

Fig. 8.2. The mechanisms through which ion bombardment acts and the critical values of the specific ion bombardment energy Φ_E and of the effective condensation rate v_D [148].

This is the equation for the distribution of "free" impurity atoms. The distribution of impurity atoms trapped by lattice defects obeys the equation

$$\frac{\partial c_f}{\partial t} = -v\, \frac{\partial c_t}{\partial x} + \frac{D}{L^2}\, c_f. \tag{8.17}$$

The solution of these equations and some computer calculations yielded the distribution of the free component of the gaseous impurities in films of finite thickness (100-300 nm) on different substrates. Satisfactory agreement was obtained between calculated and experimentally measured (by secondary ion–ion emission) concentration profiles of inert gases with energies of 100-600 eV in films of the rare earth elements.

Improvement of the mechanical and electrical properties of films by means of ion bombardment. Iskanderova, Kamardina, and Radzhabov [148] have shown that the abrasion and scratch resistance of films of Al, Ni, and Ag with thicknesses of 100-150 nm produced by vacuum evaporation and bombardment with 50-150 keV nitrogen and argon ions is reduced by a factor of 1.5-2 for doses of $1\text{-}5 \cdot 10^{17}$ cm^{-2}. Films with thicknesses of 200-600 nm become 10-50 times more resistant to abrasion after irradiation at doses of $2 \cdot 10^{15}\text{--}10^{16}$ cm^{-2}. The scratch resistance and the critical load strength of films on K-8 glass and molybdenum substrates remained at a level of 0.049-0.102 N. Increasing the ion dose to $10^{17}\text{-}10^{18}$ cm^{-2} caused damage to the films, with the formation of microscopic structure and a lowering of the mechanical durability of the film. An activation energy density of 10^{25} eV/cm^3 is required at the film–substrate interface to get an effective increase in the adhesion and mechanical durability .

Silver films with thicknesses of 250-300 nm on copper substrates produced by ion deposition at an ion energy of about 3 keV had up to *10 times the wear resistance*. The associated changes in the optical characteristics were insignificant.

Belevskii and Gusev [148] have shown that the effect of ion bombardment on the *structure and electrical properties* of niobium films produced by condensation is determined by a combination of the following ion deposition parameters: the ion energy E_i, the intensity of bombardment (more precisely, the relative intensity) n_i/n_a, the specific ion bombardment energy $\Phi_E = E_i n_i/n_a$, the effective condensation rate v_{eff}, the deposition time τ, and the specific energy E_Σ, delivered to the surface of the condensing film during this time. The "critical" or controlling parameters for ion deposition are Φ_E and v_{eff}. The different mechanisms through which ion bombardment acts on the structural characteristics and electrical properties of niobium films depend on the characteristic values of these parameters (Fig. 8.2).

For pressures on the order of 10^{-4} Pa and ion deposition of niobium films with energies of $1.6\text{-}6.4 \cdot 10^{-16}$ J (1-4 keV), the *critical value* of Φ_E is $1 \cdot 10^{-17}$ J. The *critical value* of v_{eff} is equal to the rate of displacement of the impurity quasidiffusion boundary from the substrate caused by atomic mixing of the film and substrate materials. The grain size, interplanar distances, scattering angle for the surface texture, specific electrical resistance at 300 and 10 K, the superconducting transition temperature, and the ratio of the resistances R_{300K}/R_{10K} depend on the parameters listed above.

The self-sputtering yield for niobium in the energy range $E < 1.6 \cdot 10^{-16}$ J (1 keV) as extrapolated linearly from experimental data is approximated by the analytic expression of Bogdanskii et al., and for $E > 1.6 \cdot 10^{-16}$ J, Sigmund's high-energy approximation gives values that are high (see Chapter 4). The formula

$$Y(E_i) = Y(E_i)_S - 1.4 \tag{8.18}$$

is recommended for calculations. When the average value $<Y>$ is high the effective rate of condensation of films is reduced in the region where the ion beam acts:

$$v_{eff} = [(1 - Y)\, n_i/n_a + 1]\, v_k. \tag{8.19}$$

Under these conditions an increase in n_i/n_a leads to a sharp *reduction in the superconducting transition temperature and in the resistance ratio* R_{300K}/R_{10K}, as well as to *suppression of surface texture*. This is explained by atomic mixing of the materials in the film with oxygen-containing substrates. According to Carter and Armor's quasidiffusion model for atomic mixing [148], the rate of displacement of the quasidiffusion impurity boundary away from the substrate, $v_D \sim 10^{-1}$ nm/s $\approx v_{eff}$ for the above ion deposition energy conditions. When $v_{eff} \lesssim v_D$, the niobium film is saturated with impurities from the substrate.

For 4-keV niobium ions, a pressure on the order 10^{-4} Pa, and substrate temperatures of 300 or 620 K, ion bombardment stimulates the *formation of axial tex-*

ture in very thin layers (13-26 nm) with (110) nucleation only for certain ratios of the ion bombardment intensities, n_i/n_a, and condensation rates of 0.2-0.25 or 10^{-1} nm/s, respectively. Under these conditions the superconducting transition temperature reaches a maximum of 8 K and the ratio R_{300K}/R_{10K} reaches a maximum of 2-2.8 [148].

Gusev et al. [148] have studied the interaction of 30-500 eV Cr^+ and Al^+ ion beams at fluxes of $\sim 10^{17}$ $cm^{-2} \cdot s^{-1}$ with copper substrates. The rate of deposition of Cr films at 30 eV was 1.5 μm/min, which decreases sharply with increasing energy. At 400 eV film deposition does not occur, and raising the energy further only causes sputtering of the substrate. When the ion energy is raised from 30 to 150 eV, the *grain size in the film increases* from 1 to 10 μm and the *microhardness* falls from 8 to $1.5 \cdot 10^9$ N/m^2. Cleaning the substrate with 500-eV ions for 5 s before deposition ensures good film adhesion. In the surface layer of the substrate a transition *quasidiffusion zone* is formed with a depth of 4 μm and *a concentration of atomic chromium far in excess of equilibrium* according to a phase diagram.

Entirely different deposition behavior is observed in the case of a flux of Al^+ ions. The rate of growth of the diffusion layer depends linearly on time and increases rapidly with higher energies. In the layer α and θ phases and Cu_9Al_4 are formed. The microhardness of the layers depends on the energy of the aluminum ions and varies over the range $1-1.5 \cdot 10^9$ N/m^2.

Modeling of the interaction between ions and a copper target shows that the contribution to sputtering owing to evaporation from thermal peaks during Cr^+ ion bombardment, when intermetallic compounds with Cu are not formed, is considerably greater than during bombardment with Al^+ ions, and this difference increases as the ion energy is raised.

The *growth and structure of surface diffusion layers* formed on 12Kh18N10T steel bombarded by an ionizing flux of Al vapor (98% ionized) have been studied. The ion energy was 100-500 eV and the ion flux, on the order of 10^{17} $cm^{-2} \cdot s^{-1}$. It was found that during ion bombardment the rate of growth of the layer was two orders of magnitude (!) greater than during calorizing in a medium of easily fusible alkali metals. The thickness of the diffusion layer was observed to depend linearly on the duration of irradiation. The layer is composed of the intermetallic compound phases $FeAl_3$ and Fe_2Al_5. The rate of growth of the layers increases as the ion energy is raised, and the thickness reaches 100 μm. Calculations show that the maximum possible concentration of Al in a thin surface layer may be more than twice that of Al in a stoichiometric intermetallic compound.

Nikiforova et al. [145, 148] have shown that the *structure and properties of carbon films* formed by 0-200 eV Ar^+ ion bombardment at current densities of 20-50 mA/cm^2 during condensation have a highly nonmonotonic (with several extrema) energy dependence. Electron diffraction data interpreted in terms of interference functions of small clusters of different forms of carbon together with the electronic structure of the films as derived from Fourier transforms of Auger spectra showed that when the energy of the bombarding ions lies in certain ranges, 30 ± 5, 85 ± 15, and 150 ± 10 eV, the near ordering of the films is changed. During bombardment with Cs^+ ions at energies of 0-3 keV, on the other hand, the structure of the carbon adsorbate changed only at an energy of 1.3 keV. Similar dependences have been observed in the appearance of almost resonant scattering of ions on surfaces owing to charge exchange of ions and atoms.

Diamondlike carbon films have been deposited on silicon, devitrified glass, and metal substrates by Epifanov et al. [148] using dissociation of hydrocarbons in a glow discharge with magnetron sputtering of carbon. It was found that the specific resistance of the films varied over the range 10^{10}-$5 \cdot 10^{11}$ Ω/cm.

Copper films were formed on iron and titanium at rates of 3 nm/s in a deposition system with electron beam evaporation and an accelerating voltage applied to the substrates. With an accelerating voltage of 2-10 kV and a current density on the sample of several mA/cm², copper films become *amorphous* as seen by x-rays. Films with a thickness of about 1 μm are *harder than massive samples of copper by factors of 1.5-2.7*. The hardness of the samples increases linearly with the accelerating voltage and current density at the sample. The increased hardness is related to a growth in the concentration of defects in the target. The thickness of the pseudodiffusion layer between the film and substrate is an order of magnitude greater than the theoretical maximum penetration depth for the ions into the substrate material and is practically constant over the range of ion energies that was studied. For the combination Cu^+–Fe with ion energies of up to 10 keV, the experimentally obtained thicknesses of the transition layer between the substrate and film are roughly the same at approximately 1 μm, while $10R_p \leq 50$ nm. The thickness of the transition layer is of the same order of magnitude as the height of the microscopic surface irregularities. This indicates that the topography of the substrate surface predominates over implantation processes in determining the thickness of the transition layer, since in such measurements it is the effective concentration of implanted particles, averaged over the area of a test surface that also encompasses several microscopic surface irregularities, which is determined. The thickness of the transition layer may also be increased by "smearing" owing to radiation-stimulated diffusion.

Studies have been made of the structure of films grown on the surface of NaCl crystals irradiated simultaneously by hydrocarbon molecules and 30-keV, 30 μA/cm² Ne^+ ions [268]. In the first stage of film formation, the film had the lattice structure of graphite with an anomalously large spacing of 1-2 nm, which was determined by the implantation of neon ions and residual gas particles that were completely removed by annealing at 900°C.

In the second stage of film growth, *spherical and parallelepiped-shaped formations* appeared in the film, where the latter had a grainy structure suggesting a dendritic structure. The walls of the *hollow spherulites* had the structure of the third phase of carbon–carbine. The wall thickness of the spherulites was several tens of nm. The spherulites were filled with neon, and during bombardment by Ne^+ ions they emitted radiation corresponding to atomic and molecular neon. A unit cell of the linear carbon–carbine polymer with $a = b = 0.508$ and $c = 0.78$ nm consisted of three six-atom carbon chains lying parallel to the c axis.

Thin films obtained by thermal evaporation of the metal at a pressure of $1.3 \cdot 10^{-3}$ Pa during continuous bombardment by 5-12 keV, 10-500 μA/cm^2 oxygen ions with an aluminum deposition rate of 0.2-2 nm/s may, depending on the ion dose, have the *properties of metals or dielectrics* [269] in extreme cases. Initially, dielectric islands appear in the growing film. The islands develop as the ion dose is increased, combine, and form a continuous dielectric layer. A sharp reduction in the conductivity of a layer was observed when 10^{21} O$^+$ ions·cm^{-3} had been implanted in the aluminum. When the dose was increased further, the properties of the film approached those of an Al$_2$O$_3$ film. This corresponds to a state in which 1.5-2 oxygen ions have reached the substrate for every atom of aluminum.

The dielectric islands in the first stage of deposition become charged and form local electric fields on the order of $3-8 \cdot 10^6$ V/cm. The rate of arrival of gas molecules at the surface in an electric field is increased because of the polarization of the molecules. This explains the experimental fact that films with uneven surfaces are formed at certain ion doses. On the other hand, in a high electric field autoionization of the gas occurs and sharply lowers the flux of gas molecules to the surface.

The flux of gas molecules to the surface in the presence of an electric field can be determined using the formula

$$n = n_0 \, \xi (1 - P). \qquad (8.20)$$

Here n_0 is the number of gas molecules incident on unit surface area per unit time; $(1 - P)$ is the probability that a gas atom is able to avoid ionization; and ξ is the enhancement in the flux of molecules on the surface owing to an electric field of strength E_0 given by

$$\xi = \frac{\mu E_0}{kT} + \left(\frac{\pi \alpha E_0^2}{2kT} \right)^{1/2} \mathrm{erf} \left[\left(\frac{\alpha E_0^2}{2kT} \right)^{1/2} \right], \qquad (8.21)$$

where μ is the dipole moment of the gas and α is its polarizability. If $(\alpha E_0^2/2kT)^{1/2} > 2$, then $\mathrm{erf}(\alpha E_0^2/2kT)^{1/2} \approx 1$ and

$$\ln \frac{n}{n_0} = \ln \left\{ \mu \frac{E_0}{kT} + \left[\frac{\pi \left(\frac{1}{2} \alpha E_0^2 \right)}{kT} \right]^{1/2} \right\}$$
$$- \frac{R\nu}{2a} \left(\frac{E_0 M}{2a} \right)^{1/2} x_0^{-1/2} \exp(-x_0), \qquad (8.22)$$

where the first term on the right is the enhancement in the molecular influx owing to polarization and the second is the attenuation of the influx owing to autoionization (Fig. 8.3).

Increasing the concentration of oxygen atoms implanted in a growing aluminum film converts it into a dielectric, Al$_2$O$_3$ (Fig. 8.4), and ensures that the film will grow evenly with a uniform surface.

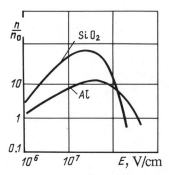

Fig. 8.3. The calculated flux of molecules onto a substrate as a function of the electric field strength for $T = 500$ K and $R = 50$ nm [269].

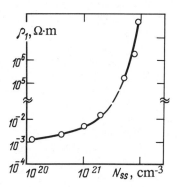

Fig. 8.4. The dependence of the specific resistance of aluminum films on the amount of implanted oxygen ions [269].

Epitaxial growth in the thickness of antimony films on chips of NaCl monocrystal has been achieved by thermal deposition of antimony during irradiation with a flux of 400-eV, 3-5 µA/cm² argon ions [270]. The sputtering system was pumped down to a pressure of $1.3 \cdot 10^{-4}$ Pa, with forced freezing out of oil vapors and heating of the substrate to 300°C in order to outgas it. Epitaxial growth of films with one orientation (1012) was observed at substrate temperatures of 150-300°C. In the absence of ion bombardment, at least three orientations were observed, and epitaxial growth began at a temperature of 230°C. On cubic crystals a further increase in the thickness of the condensate layer leads to the formation over the entire surface of a continuous monocrystalline film with a high degree of orientation.

Experimental and theoretical studies of the condensation of Zn on Cu_2O accompanied by irradiation by 100-300 eV Ar^+ and Ne^+ ions with a current density of up to 10 µA/cm² demonstrate the possibility of *controlling the formation and structure of thin films* [271]. Ion bombardment is of particular interest for stimulating various film formation processes because the ion–atom interaction cross sections are several orders of magnitude greater than those for electron–atom and photon–atom processes.

Ion bombardment changes the pre-exponential factor in the formula for the critical pressure P_c:

$$P_c \sim (1/\sigma) \exp\left[-U_0/(kT)\right], \tag{8.23}$$

where σ is the cross section for capture of an atom and U_0 is the binding energy of an atom on the surface. At the same time, as a comparison of the functions $\log(P/P_0) = f(1/T)$ shows, irradiation has no effect on the binding energy between the film and substrate.

The experimental dependence of P_c on the energy flux of an ion beam at the surface (mW/cm^2), as well as the function $P_c = f(j)$, are given by the same formula

$$P_c = P_0/(1 + \sqrt{\beta j E}), \tag{8.24}$$

that is, $N \propto 1 + \sqrt{\beta j E}$. This formula is analogous to that for the equilibrium concentration of point defects in the surface layer of Cu_2O semiconductor,

$$V = V_0/(1 + \sqrt{\beta j E}), \tag{8.25}$$

where V_0 and V are the concentrations of vacancies in the absence of ion bombardment and with it and β is the recombination coefficient for pairs of defects.

The kinetic equations together with the assumption that the condensation centers and point defects are coupled yield the expression

$$P = P_0/(1 + \sqrt{\beta j E})^{1/(n+1)}, \tag{8.26}$$

where n denotes the dimensions of the critical nuclei and P_0 is the critical pressure in the absence of ion bombardment.

Optical interferometry can be used for studying film coatings on surfaces. It is based on the relationship between the coating factor of the surface and the intensity of the reflected light:

$$R = \sqrt{J_1}\,(1-x)\,\exp(i\,\varphi_1) + \sqrt{J_2}\;x \exp(i\,\varphi_2), \tag{8.27}$$

where R is the amplitude of the reflected wave; $\sqrt{J_1}$ and $\sqrt{J_2}$ are the amplitudes of the reflected waves from the substrate and the film surface, respectively; φ_1 and φ_2 are the corresponding phases of the reflected waves; and x is the (fraction of the) surface area coated by the film and $(1 - x)$ is the uncoated fraction. Measurements made in this way agree with electron microscopic data. These measurements have shown that, first, *ion bombardment accelerates the coating of the surface by condensate*, second, *ion bombardment reduces the size of the islands*, thereby increasing the continuity of the film, and third, the changes are more distinct when ion bombardment begins earlier.

A modified Kolmogorov equation can be used to describe the variation in the degree of coating. Its solution for small t gives

$$x = (\gamma t/2)^2, \tag{8.28}$$

where

$$\gamma = (a^2 SPI/2)^{1/2}. \tag{8.29}$$

Here a^2 is the area of the surface occupied by condensed atoms in islands; S is the maximum area of an island ($S \sim 1/\sqrt{jE}$); P is the flux of vapor atoms onto the surface; and I is the rate of nucleus formation.

In the early stages ($t < 50$ s) Eq. (8.29) agrees well with the data from all experiments. For 400-eV argon ions

$$I \sim j^{5/2} \sim N\Gamma, \tag{8.30}$$

where j is the ion current density and Γ is the flux of atoms to the condensation centers. Since $N \sim j^{1/2}$, we have $\Gamma \sim j^2$.

In the early (critical) stages of formation of a film, therefore, the main effect of ion bombardment is related to an increase in the concentration of condensation centers, since $N \sim \sqrt{jE}$, while the large number of condensation centers ensures that any structure is of small size. The kinetics of further film growth is determined by the increased mobility of atoms owing to ion irradiation. The rate of film growth is proportional to j^2. Island migration is also activated and stimulates the coating process, which acts on the mechanism for growth of the film and its structure. All of this makes it possible to obtain films of high quality, as good as monocrystalline.

The effect of ion bombardment during film growth has been examined in the course of research on the following topics: the superconducting properties of niobium with Kr^+ ion energies ≤ 500 eV; the structure of titanium layers on molybdenum over the temperature range 115-970°C for growth rates of 0.25-3 μm/min and ion energies of up to 10 keV; the deposition rate, adhesion, and structure of layers of copper, gold, platinum, and chromium deposited on aluminum oxide, quartz glass, steel, and brass during irradiation by argon ions in a discharge plasma at a pressure of $4 \cdot 10^{-2}$ Pa; and the adhesion and interaction chemistry of selenium with brass and copper substrates.

Ion plating has been used to make *telephone relay contacts* with a low contact resistance and a long operating lifetime [272]. The methods used up to that time, electrodeposition, electrolytic deposition, vacuum evaporation, vacuum deposition, deposition of metallic resins, and plasma deposition, did not provide the required properties. Ion deposition seemed better than sputtering since harder films with more resistance to crack formation could be obtained. The best conditions for ion plating are: argon pressure 2.66 Pa, substrate potential −4 kV, and deposition rate 10 nm/s. The problem of coating reproducibility was solved by mounting a large number of contacts in a holder in the shape of a "coliseum." The main advantage of this holder is the uniformity of the glow discharge density and the very uniform thickness of the coating, with coating only at the contact tip. A second advantage of this holder is a high degree of automation, since it can be rotated until all 1100 contacts have been plated with a precision coating of metal.

8.4. Vacuum–Plasma Film Deposition
(A Method for Condensation by Ion Bombardment)

Basic research on the physics of high temperature plasmas for controlled thermonuclear fusion at the Kharkov Physicotechnical Institute of the Academy of Sciences of the Ukrainian SSR has led to the development of high-energy plasma technology [129].

A vacuum arc in cathode mode and plasma accelerators are used to obtain a flux of metallic plasma. The high current density in the cathode spots (1-10 MA/cm^2) and surface energy flux (10-100 MW/cm^2) cause intense vaporization of the cathode and the generation of a plasma jet consisting of cathode erosion products. The degree of ionization of the refractory metals is as high as 80-100%. The bulk are doubly charged ions with an average energy on the order of 100 eV. The ion current reaches 10 A or more and constitutes about 10% of the discharge current. The energy efficiency of plasma generation may reach 20-30%, and the rate of deposition of condensate, up to 0.5 µm/min. Deposition of plasma condensates is carried out in "Bulat"-type systems. In order to obtain chemical compounds (nitrides, carbides) in a vacuum vessel with a base pressure on the order of $6.6 \cdot 10^{-4}$ Pa, the appropriate gas is fed into the vessel.

The ionization state of the material and the interaction of the high-velocity ions with a solid surface make it possible to obtain coatings with superior physical and mechanical properties, i.e., that are superhard, wear resistant, etc. Since the coatings are formed directly from an ion beam, they are subjected to ion processing over their entire depth. It is possible to specially change the phase composition, structure, and properties of plasma coatings by varying the composition and energy of the ion beam. Coatings consisting of the pure initial vaporized materials, their solid solutions, compounds, and heterogeneous alloys are thus obtained. Multilayer compositions with programmed changes in the contents of different components have been fabricated.

Accelerated ion beams with a high current density have been used for highly efficient initial cleaning of sample surfaces by sputtering. The resulting high adhesion cannot be obtained by any other existing coating technique.

Coatings of the nitrides of molybdenum (Mo_2N) and titanium (TiN) had maximum microhardnesses of $4.3 \cdot 10^{10}$ and $5.9 \cdot 10^{10}$ N/m^2, respectively [122]. For an arc current of 120 A with a magnetic field of $1.6\text{-}2.4 \cdot 10^3$ A/m at the anode for focusing the plasma jet, the maximum rate of growth of molybdenum condensate was 200 µm/h when the cathode–substrate distance was 300 mm. The titanium ion beam current was 7-8% of the arc current.

TABLE 8.4. Characteristics of Different Methods of Film Deposition [152]

Characteristic	Vacuum arc burning in microspots on a consumable cathode	Vacuum arc with a distributed discharge on a consumable cathode	Vacuum arc with a consumable anode and permanent cathode	Magnetron discharge (in crossed E and B fields)	High-voltage electron beam in vacuum	Ion beams in vacuum
Power, W	$5\cdot10^3$	$5\cdot10^3$	$5\cdot10^3$	$5\cdot10^3$	$5\cdot10^3$	0.2-10^7
Operating pressure, Pa	$1.3\cdot10^{-3}$	$1.3\cdot10^{-3}$	$1.3\cdot10^{-3}$	$1.3\cdot10^{-1}$	$1.3\cdot10^{-3}$	$1.3\,(10^{-3}$-$10^{-4})$
Voltage, V	10-30	10-20	10-50	300-800	10^4	20-10^5
Current, A	150-500	250-500	100-300	10	0.5	10^{-3}-10^2
Current density, A/cm² at cathode (in spots)	10^5-10^6	5-100	1-10	0.003-0.3	0.5-10	10^{-3}-10^{-1}
at anode	DD	DD	10^1-10^3	DD	1-10	10^{-3}-10^2
Electron emission mechanism	TE	AE	AE	PE	TE	TE
Consumable electrode and mechanism for its destruction	K_e	K_e	A_e	K_s	A_e	$K_{e,s}$
Energy source	IB	IB	EB	IB	EB	IB
Rate of consumption of electrode, g/min	0.3-1.5	0.5-5	0.1-10	0.1-0.7	0.4-2.1	0.4-2.1
Particle energy, eV	10-100	1-2	0.1-1	10-20	0.1-0.3	1-100
Phase composition	I, At, Md	I	I, At	At, I	At	I, At
Degree of ionization, %	10-90	100	5-100	10-20	—	10-100
Energy cost per unit mass, J/g;	10^5-$6\cdot10^5$	$2\cdot10^4$-$3\cdot10^5$	$5\cdot10^3$-10	$5\cdot10^5$-$5\cdot10^6$	$5\cdot10^3$-10^5	10^6
per atom, eV/atom	300-860	40-300	5-100	200-2000	5-100	30-$5\cdot10^3$

Note. DD, distributed discharge; TE, thermionic emission; AE, anomalous emission; PE, potential (field) emission; K, cathode; anode; e, evaporation; s, sputtering; IB, ion beam; EB, electron beam; I, ions; At, atoms; Md, microscopic droplets.

The nitrides of titanium and molybdenum are used to strengthen instruments made of hard alloys and fast cutting and instrument steels. Laboratory and industrial tests of an instrument that had been hardened in this way showed that its durability for working structural steels and alloys, cast iron, and nonferrous metals had been increased by a factor of 2-8. Specialized setups equipped with "Bulat-2" and "Bulat-3" systems for hardening cutting tools have been working in a number of factories since 1975 [129].

High-energy ions produce pressures as high as hundreds of kilobars in a target, and the thermal energy that is released may correspond to temperatures of several thousand degrees. By removing heat from these pieces through cooling of the substrate it is possible to quench metastable crystalline formations, including diamonds. Carbon ion energies of several tens of eV are sufficient to obtain the latter. The artificial diamond coatings have a density of 2.9 g/cm^3, specific resistivity of 10^8 Ω/cm, and microhardness of $9.8 \cdot 10^{10}$ N/m^2.

The energy of the ions has a significant effect on the microstructure and macrostresses that develop on the coatings. Thus, for an ion current density of about 10 mA/cm^2, when the average energy of the molybdenum ions was changed from 200 to 720 eV and the condensation rate from 16 to 10 μm/h, the temperature of substrates made of 40Kh steel changed from 90 to 150°C. As the ion energy was increased, the width of columnar crystallites increased from 0.2 to 0.5 μm. For an ion energy of 720 eV the microhardness in a 20-μm-thick molybdenum coating was $5.7 \cdot 10^9$ N/m^2 and a minimum value of the microhardness of $2.4 \cdot 10^9$ N/m^2 was observed in the steel at a depth of 4-16 μm [273]. This is explained by the development of vacancy porosity, since the steel–coating interface is an effective sink for vacancies from the steel and molybdenum.

The microhardness of the condensates is raised to $(5.2 \pm 0.20) \cdot 10^9$ N/m^2 compared with the $3.7 \cdot 10^9$ N/m^2 of massive molybdenum samples. This is a consequence of the change in its phase composition, the formation of fine-grained structure, and a high level of micro- and macrostresses. When the average ion energy is changed from 200 to 720 eV, the macrostresses change from –15.1 to –28.1 in σ units of $5 \cdot 10^{-7}$ N/m^2.

Protective coatings have been used in dentistry for metal dentures. Vacuum–plasma coatings are 14 times more durable and longer lasting than gold and are 2-3 times more resistant to wear than steel crowns [274]. This method of depositing protective coatings was developed under the leadership of M.A. Napadov and has been introduced in 13 dental clinics in Kharkov.

Dorodnov and colleagues [121, 152] have discovered two new types of stationary vacuum arc: an arc with an extended discharge on a hot consumable cathode and an arc with a permanent hollow cathode that burns in the vapor of the anode. The first type is realized when the surface energy flux is sufficient to produce a high temperature on the cathode surface, typically above 10^3 K. On thermally isolated cathodes this type of arc exists for moderate current densities at the cathode, above some critical value of 10-100 A/cm^2 which is a function of the thermal properties of the cathode material. The fluxes of metal plasma are fully ionized and contain no microscopic droplet phase. The second type of vacuum arc is realized only in an electrode system with a certain geometry. The anode must be located inside a hollow cathode made of a material with high emissivity characteristics. The principal characteristics of the different coating techniques are shown in Table 8.4.

The PUSK 77-1 system, with an end-mounted Hall plasma source, has been developed and placed in industrial use in the Soviet Union for making wear-resistant thin coatings of titanium nitride [121, 122, 275]. The coating material is formed by combining titanium which enters the source from a consumable cold cathode with nitrogen which fills the vacuum vessel at a pressure 0.12 Pa. The growth rate for coatings on a surface mounted perpendicular to the plasma flow a distance of 300 mm from the cathode is 1 μm/min. The microhardness of the coating is 1.96·10^{10} N/m^2. The basic parameters of the source are a voltage of 30 V and a current of 140 A.

Penning sources with extended heating regions developed at the Kurchatov Atomic Energy Institute and at the Bauman Higher Technical School have been used in space technology and in the electronic, microelectronic, optical, and metallurgical industries [122-124, 126-128, 169, 175].

Pulsed plasma sources have been shown by Osadin and Shapovalov [122] to be capable of reaching high instantaneous production and condensation rates of 0.1-10 μm/s with an average condensation rate of 5·10^{-2} μm/s. The power of their pulsed plasma source was 1 kW with a pulse repetition rate of 300 s^{-1} and an ion current density of about 1 A/cm^2. A pulsed plasma source can be used to deposit alloy films which maintain the initial component composition. With the aid of a 600-J capacitor bank, films with thicknesses of several hundred nanometers were deposited over a series of discharges with repetition rates of 1/40 s^{-1}. Depending on the discharge energy (54-600 J), the increase in the deposited mass of film was from 0.02 to 0.5 μg/cm^2 per discharge. A Cu–Zn (10%) alloy was deposited on a substrate using a pulsed plasma source while maintaining the zinc content in the alloy to within 1%.

The technology of thin film deposition occupies a prominent place in the production of electronic and optical components. The editors of the journal "Elektronika" state that the volume of trade in electronic components in the USA, Western Europe, and Japan in 1976, 1983, and 1984 was 83, 155, and 184 billion dollars [276]. In 1984 the US expenditure on electronics reached $184 billion or about 20% of the federal budget, in which planned military expenditures totaled $240 billion. The fraction of plasma technology in electronics manufacturing was about 20% in 1976 and had increased to 40% by 1980. Up to now, however, the main methods for producing thin films have been evaporation and sputtering. Ion deposition and ion implantation and plating are developing along with them. New possibilities for producing textured coatings under ion bombardment conditions in glow discharges have also been discovered.

TABLE 8.5. The Effective Sputtering Rate (v_R) and Sputtering Yield (Y_R) of Tungsten Deposited from Tungsten Chloride in a Glow Discharge [117]

Pressure of tungsten chloride, Pa	Deposition temperature, K	Discharge voltage, V	Current density, A/m²	v_R, μm/s	Y_R, atoms/ion
67	1470	1200	300	0.16	0.20
67	2270	1250	1250	1.00	0.30
670	2270	870	1400	0.30	0.01

8.5. Production of Textured Tungsten Layers in Glow Discharges

Babad-Zakhryapin and Kuznetsov [117, 117a,b] have generalized the results of research on the production of textured high-temperature coatings and improving the properties of metal surfaces by ion bombardment under glow discharge conditions. The interaction of ions with the surface leads to formation of activated regions because the number of free bonds is increased by the removal of impurities on the substrate surface and of atoms from the substrate itself. The concentration of point defects increases, more atomic migration occurs, and more nuclei are formed, so that the rate of formation of nuclei for film growth increases. Ion bombardment causes formation of a limited number of growth directions and crystallographic planes. When films are produced by the decomposition of chemical compounds under ion bombardment conditions, the substrate is heated, and dissociation, activation of chemical reactions, and ion sputtering take place. Metal coatings (Nb, Mo, Ta, W, etc.) are deposited from halogens via reactions of the type

$$MeR_x \xrightarrow{q^+} MeR_y + R \tag{8.31}$$

and

$$MeR_y \xrightarrow{q^+} Me + MeR_x, \tag{8.32}$$

where Me is the metal or semiconductor, R is hydrogen, a halogen, or a radical, and q^+ are positive ions.

The deposition of oxide coatings (Al_2O_3, SiO_2, Ta_2O, Nb_2O_5) from chemical compounds (metal organic, silicon organic, chlorides, fluorides, etc.) in a glow discharge takes place through the reaction

$$MeR_x + O_2 \xrightarrow{q^+} MeO_y + R. \tag{8.33}$$

The balance equation for the adsorbed reagents can be written in the form

$$\frac{dh}{d\tau} = -\sigma j\, h/q_i - Y_R\, j\, M\, (q_i\, \rho N_A) + (h_0 - h)/\tau_s, \tag{8.34}$$

where h is the average effective thickness of the adsorbed layer; τ is the instantaneous time; σ is the effective interaction cross section for the ions with molecules adsorbed on the cathode; q_i is the ionic charge; Y_R is the effective sputtering yield; ρ and M are the density and molecular mass, respectively, of the adsorbed material; N_A is Avogadro's number; h_0 is the value of h without ion bombardment; and τ_s is the lifetime of the adsorbed molecules.

The second term in Eq. (8.34) characterizes the rate at which the thickness of the adsorbed layer is reduced by cathode sputtering of adsorbed molecules.

In the steady state, when

$$\tau \gg (\sigma j/q_i + 1/\tau_s)^{-1}, \tag{8.35}$$

$$h = [h_0 - Y_R j M \tau_s / (q_i \rho N_A)] (\sigma j \tau_s / q_i + 1). \tag{8.36}$$

The rate of deposition of the layer resulting from ion bombardment alone is given by

$$v_i = \frac{M_0 \rho \sigma j}{M \rho_0 q_i} \left[h_0 - \frac{Y_R M j \tau_s}{q_i \rho N_A} \right] \bigg/ \left(\frac{\sigma j \tau_s}{q_i} + 1 \right). \tag{8.37}$$

It is clear from this equation that for sufficiently large j the deposition rate becomes negative, i.e., sputtering predominates over film growth.

Table 8.5 lists the parameters for tungsten coating in glow discharges.

A need for textured tungsten coatings has arisen in connection with research and development work on thermionic converters of thermal into electrical energy [110]. These coatings have a maximum electron work function in vacuum and a minimum when covered with a monolayer of cesium. Limited supplies of tungsten, however, require that it be used sparingly as a thin film coating on metals and other materials with lower melting points or high sputtering yields. The latter point is important in the construction of high-voltage electrodes for ion-optical systems in megawatt ion beams and neutral injectors for thermonuclear experiments [277]. The use of tungsten coatings, including textured coatings, makes it possible to increase the operating lifetime of ion sources by several times and to improve the electrical properties of thermionic converters.

The making of coatings in glow discharges involves a number of complex processes in which fast and bulk plasma electrons and electromagnetic radiation participate, along with the ions. In this regard, ion-beam film deposition takes place under simpler physical conditions, although more complicated equipment is used.

8.6. Film Deposition by Ionized Clusters

During sputtering by heavy ions, multiatomic or cluster ions of the substrate material are formed, along with atoms, ions, molecules, and molecular compounds. For example, as noted in Section 4.6, during sputtering of a number of substances (Al, Si, V, Co, Cu, Ag, Au, Nb, Ta, Mo, and W) by 8.5-keV Kr^+ and Xe^+ ions,

N. Kh. Dzhemilev et al. [89] recorded the following ions in mass spectra: Al_n^+ ($n = 1$-17), Si_n^+ ($n = 1$-12), V_n^+ ($n = 1$-16), Co_n^+ ($n = 1$-15), Cu_n^+ ($n = 1$-39), Ag_n^+ ($n = 1$-41), Au_n^+ ($n = 1$-21), Nb_n^+ ($n = 1$-17), Ta_n^+ ($n = 1$-13), Mo_n^+ ($n = 1$-17), and W_n^+ ($n = 1$-17). The targets were cleaned by heating at temperatures of 0.7-0.8 times the melting temperature and pressures of $1.3 \cdot 10^{-7}$ Pa for 20-40 h, as well as by bombardment with a 3 mA/cm^2 scanning ion beam.

A number of models have been proposed for the formation of clusters during sputtering of solids, including the cascade-hydrodynamic model. Bitenskii and Parilis [89] found the average value of the sputtering yield to be

$$Y = (1 - \overline{Q}) \, Y_l + \overline{Q} Y_\omega, \tag{8.38}$$

where $\overline{Q}$ is the probability of formation of a cascade with a high density of collisions, and Y_l and Y_ω are the sputtering yields obtained from the linear collision theory of Sigmund and the shock wave theory of Kitazoe and Yamamura (see Chapter 4).

Another way of obtaining ionized clusters is to cool the gas to the condensation temperature by adiabatic expansion in a nozzle that enters a vacuum, where the gas condenses into clusters. A beam of neutral clusters can be ionized by, for example, electron collisions in a Bayard–Alpert ionization tube. This method has been used to obtain a beam of nitrogen clusters with an equivalent current of 28.6 mA [278]. With an electron current of 10 mA and energy of 50 eV an extraction efficiency of 65% was obtained. An average-sized cluster at a pressure of $8.8 \cdot 10^4$ Pa and temperature of 77.3 K contained $2 \cdot 10^5$ molecules. Injecting hydrogen clusters is one way of heating the plasma and fueling a steady-state thermonuclear reactor.

Clusters of Cu, Si, ZnS, and Pb with 10^2-10^3 atoms have been obtained by producing vapors of the metals and semiconductors through heating them in a crucible and letting the vapor expand adiabatically into a vacuum [279]. For ion currents of 1-60 μA the rate of film deposition varied from several tens of nanometers to several microns per minute. Good quality films were obtained, and the deposition system is fairly simple and suitable for industrial applications, for example, in preparing semiconducting p- and n-type layers [280].

8.7 A Source of Sputtered Atoms for Laser Isotope Separation (of Uranium and Other Elements)

Laser isotope separation is based on the existence of a tiny difference between the emission and absorption spectra of both the isotopes of a single element and of the compounds of these isotopes. Narrow-band light sources, i.e., lasers, can in principle selectively excite the isotopes in mixtures of atoms and molecules. Since the physical and chemical properties of excited atoms and molecules are substantially different from those of unexcited ones, it is possible to separate the isotopes using one or more of these properties [281].

Laser enrichment systems require less electrical energy, and the cost of building them is decreasing with time. Estimates show that the cost for the Western countries of producing the enriched uranium needed to operate nuclear power plants through the year 2000 would be reduced by roughly $100 billion.

The source of atomic vapor for laser isotope separation must meet a number of requirements, including: the flux of neutral atoms, their velocity, the qualitative composition of the atomic vapor (presence of parasitic ions), the populations of the working levels in the separation zone, and the design features of the atomic vapor source. Experimental studies of ways of obtaining neutral atoms of the refractory and chemically active rare-earth element Gd have shown that electron-beam vaporization and cathode sputtering are promising. A hollow cathode source operating in a krypton atmosphere at a pressure of 6.6-$9.3 \cdot 10^{-2}$ Pa with a current of 100 mA was used as a source. At a distance of 50 mm from the cathode the number density of Gd ions was $4 \cdot 10^{6}$ cm^{-3} and that of Gd atoms, 10^{4} times greater. The flux of atoms was as high as $1 \cdot 10^{15}$ cm$^{-2} \cdot$s^{-1}. The relative population of the ground $a^{9}D_{5}$ (999 cm^{-1}) term was roughly six times greater than the population of these levels with thermal heating to 2000°C.

Chapter 9

ION-BEAM WELDING, MICROANALYSIS, AND MICROMACHINING OF SOLIDS WITH ION PROBES

9.1 Ion-Beam Welding

High current-density heavy-ion beams. It is much easier to produce electron beams with the high surface energy flux needed for welding than ion beams. Nevertheless, in a number of cases welding with an ion beam may be more suitable, since an ion beam can be used simultaneously to clean the surface. It is less sensitive to external magnetic fields than an electron beam. Almost no x-rays are produced during ion-beam welding [283]. In a comparison of electron or laser beam welding with possible ion-beam welding, however, attention should first be turned to a fundamental feature of the latter: only an ion beam can be used directly to introduce additives for changing the composition of the welding seam. In the way of practical realization of this idea, on the other hand, lies the difficulty of creating a beam of heavy ions with a high current and high current density

$$j_+ = \eta \, n_a \, h \, e \, v \, / \, (\xi l),\tag{9.1}$$

where η is the relative number of impurity atoms needed to give the seam the required properties, n_a is the number of atoms/cm^3 of the metal to be welded, h is the seam depth, e is the electronic charge, v is the welding speed, ξ is the absorption coefficient for the incident ions, and l is the dimension of the beam in the direction of motion. Taking as an example $\eta = 1\%$, $n_a = 6 \cdot 10^{22}$ atoms/cm^2, $h = 0.5$ cm, $\xi = 1$, $v = 0.1$-1 cm/s, and $l = 0.2$ cm, we obtain $j_+ = 24$-240 A/cm^2, which gives a beam current $I_+ = 1$-10 A for a beam with a cross-sectional area of 0.04 cm^2. An analysis shows that it is extremely difficult, for a number of reasons, to produce a steady-state beam of heavy ions with these parameters in the customary way, i.e., by separating the charges near the plasma boundary of a source and then accelerating, focusing, and transporting the ions to the piece to be welded. For example, the high potentials needed for this method would lead to disproportionately high beam

187

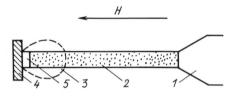

Fig. 9.1. An apparatus for ion-beam welding: (1) discharge vessel; (2) plasma column transported by the magnetic field into a vacuum; (3) luminous flare during welding; (4) work piece; (5) positive space charge layer.

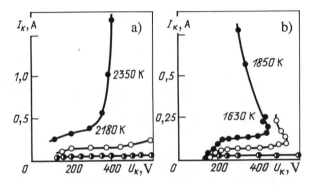

Fig. 9.2. The volt–ampere characteristics of the work piece as an ion collector for different plasma densities: (a) molybdenum collector; (b) titanium collector.

powers. Thus, we turn to another way of obtaining ion beams for welding: accelerating the ions extracted from the plasma in the Debye layer between the plasma and the object to be welded [284].

Let us suppose that a column of dense plasma with a diameter on the order of 1 mm (Fig. 9.1) is transported with the aid of a magnetic field from a gaseous-discharge device into a vacuum region with a residual gas pressure on the order of 10^{-3} Pa that contains the metal to be welded. The work piece is held at a negative potential of up to 1 kV relative to the plasma. It was possible in this way to obtain an on-axis ion current density of up to 100 A/cm^2 with a surface energy flux of up to 10^5 W/cm^2 and a total current of several amperes. The form of the volt–ampere character-istics of the work piece as an ion collector (Fig. 9.2) is explained as follows [285]: When it leaves the plasma and strikes the collector, a gas ion knocks out γ_1 electrons and Y_1 metal atoms, each of which has a high probability β of becoming a positive ion as it penetrates the dense plasma with its fairly high electron temperature. The positive ion returns to the collector and knocks out γ_2 electrons and Y_s

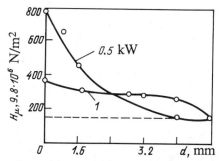

Fig. 9.3. The variation with depth of the microhardness in a transverse cross section of a sample for two values of the ion energy flux (dashed curve is the microhardness distribution before processing).

atoms (Y_s is the self-sputtering yield). This sequence of processes is repeated, and the ion current density is equal to

$$j_+ = j_{+0}(1 + \gamma_1) + j_{+0} Y_1 \beta (1 + \gamma_2)/(1 - \beta Y_s) \qquad (9.2)$$

and should increase rapidly as $\beta Y_s \to 1$. This avalanche growth in the ion flux is caused primarily by the existence of a strong external ionizer (a plasma with $\beta \to 1$) rather than by ionization of the atoms knocked out of the collector by electrons. Ion "multiplication" should be observed when $\beta Y_s = (Y_s n_e < \sigma_i v_e > \lambda_i)/v_a \geq 1$, where v_a is the average velocity of the sputtered atoms and λ_i is their mean free path in the plasma. In the case of a radially bounded plasma, $\lambda_i \approx R$, where R is the radius of the plasma column. For example, when $n_e = n_+ = 10^{14}$ cm^{-3}, $v_e = 2 \cdot 10^8$ cm/s, $\sigma_i = 10^{-16}$ cm^2, and $v_a = 10^5$ cm/s, the value of β reaches its maximum (1) by the time $R \approx 0.1$ cm. As for Y_s, at an energy of about 500 eV it may be close to or even greater than 1, especially when the collector temperature is close to the melting point (see Chapter 4). Thus, welding takes place with strong cathode sputtering, ion multiplication, and an intensely luminous flare. It is important to emphasize that all of this occurs without a transition to an arc discharge with a small voltage drop and a cathode spot.

Deep alloying of welding seams. We shall give two examples illustrating welding with simultaneous introduction of alloying additives. Figure 9.3 shows the microhardness distribution with depth in a sheet of titanium iodide after it was welded with a carbon ion beam extracted from a CO_2 discharge plasma [284]. The change in the microhardness indicates that carbon was introduced over the entire seam thickness, and the observed deep implantation is evidence of intensive convection in the welding puddle.

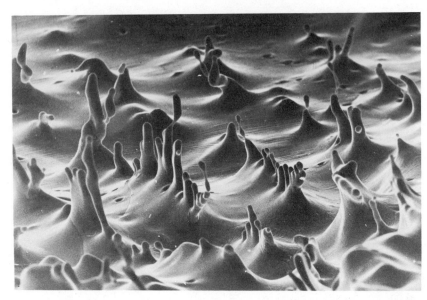

Fig. 9.4. The panorama of "frozen" capillary waves on the back side of a weld puddle.

In another experiment [286] a sheet of 3-mm-thick stainless steel was welded with a beam of titanium and nitrogen ions (Table 9.1). The concentration of implanted titanium was determined from the characteristic x-ray emission produced by an electron probe. The table shows that the ion flux made it possible to introduce more than 1% of titanium into the seam during the welding period.

The ion-acoustic instability and capillary waves. One other important property of the ion flux has been observed during research on this topic. The following facts were first established [287]:

(1) An ion-acoustic instability with a frequency of $\omega = (2\pi/L)(\chi T_e/M)^{1/2}$ develops in the plasma column between the gaseous-discharge vessel and the work piece, where L is the length of the column, χ is the Boltzmann constant, T_e is the plasma electron temperature, and M is the ion mass. The calculated frequency coincides with the experimentally observed frequency of deep modulations in the ion flux. For a hydrogen plasma, this frequency is on the order of 10^6 Hz.

(2) The recoil pressure of the atoms sputtered by the intense ion flux was substantial, and the modulation in the ion flux caused modulation of the recoil pressure and, therefore, excitation of intense ultrasonic oscillations in the welding puddle at this frequency. The high intensity of the oscillations is confirmed, in particular, by the nonlinear capillary waves observed on the liquid-metal surface (Fig. 9.4). Large-amplitude capillary waves have been observed with a scanning electron microscope from the back side of a seam, where there is no plasma or electric field which might cause other types of instabilities on the liquid-metal surface. This panorama develops in the following way:

Fig. 9.5. The panorama of "frozen" capillary waves on the surface of a weld puddle facing an ion flux.

TABLE 9.1. The Titanium Content at Different Depths in Stainless Steel before and after Ion-Beam Welding

Depth, mm	Titanium concentration, %	
	before welding	in the welding seam
1.0	0.2	2.1
1.5	0.2	1.2
2.5	0.2	1.4

Capillary waves with a frequency of $\omega = (\sigma k^3/\rho)^{1/2}$ can propagate on the surface of the liquid metal, where σ is the surface tension, $k = 2\pi/\lambda$ is the wave number, λ is the wavelength, and ρ is the density of the metal. The amplitude of these oscillations, as that of oscillations in a solid, is small. However, for a sufficiently high recoil pressure from the sputtered atoms and a certain frequency of modulation of this pressure ($f = \omega/\pi$), parametric excitation of nonlinear capillary waves occurs [287]. If the ion flux is cut off, then the oscillations damp out in the deep part of the melt long before the metal solidifies, and in this zone no noticeable protrusions are recorded on either the front or back. Some fraction of the metal is thrown out of the weld puddle onto the "cold shore" where the capillary waves may be "frozen." This means that the metal on the periphery of the melt zone solidifies before the waves are damped, and their "panorama" is recorded there. Capillary waves excited by ultrasound are recorded in pure form only on the back side of the metal. On the front side there is an electric field E which can modify the panorama through the so-called Tonks–Frenkel instability (Fig. 9.5). Under these conditions this instability involves the following: a strong field E produces a "negative pressure" (directed inward) on the metal surface. The initial perturbation of the surface

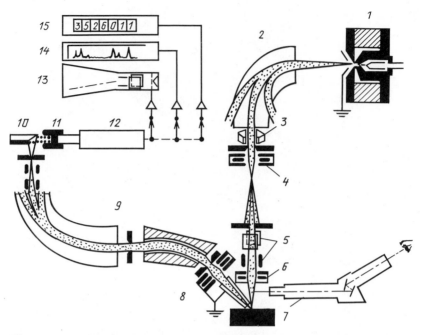

Fig. 9.6. An analyzer system: (1) duoplasmatron; (2) mass separator; (3) alignment plates; (4) einzel lens; (5) deflection plates; (6) short-focus lens; (7) device for observing the surface of the analysis target; (8) lens; (9) double-focus mass spectrometer; (10) electron emitter; (11) scintillator; (12) photomultiplier; (13) cathode ray tube; (14) recorder; (15) counter.

caused by the ultrasound is amplified by the negative pressure, and this in turn leads to a growth in the local field, to further development of the perturbation, etc.

Ultrasonic processing of welding seams. In comparing Figs. 9.4 and 9.5 it may be noted that, despite the intrinsic similarities in the structures of the panoramas induced by ultrasound, the front and back sides have different peak shapes. Where there is an electric field the peaks have a smaller radius of curvature and are also "frozen" at the moment a droplet breaks away.

It has thus been established that the ion flux used in welding can not only introduce the required elements into the welding zone, but can also subject the welding zone to intense ultrasonic processing. This can intensify degassing of the metal and also, for example, speed up the reduction in the grain size. This is consistent with experiments [288] and extends the range of possible applications for this method of ion welding.

9.2. Ion Probes for Microanalysis and Micromachining of Solids

Ion submicron probes. Among the standard methods for obtaining information on the composition, structure, and properties of solids, those based on the use of charged-particle beams occupy a prominent place. Electron microscopic studies of surfaces and local surface analysis in terms of characteristic x-ray emission produced by fast electrons are well known. There is also a variety of methods for diagnostics of solids using ion beams [138-140]. The use of well-focused, dense ion beams (ion submicron probes) with relatively low energies (on the order of tens of keV) has attracted the particular attention of specialists in various fields (physicists, technologists, geologists, archaeologists). The ability to make submicron ion probes opens up wide perspectives for their use in scientific and applied research. Since a number of reviews have been published [172-174], in this section we shall dwell primarily on the latest achievements in the production of such beams with liquid-metal ion emitters.

The principle of ion-beam microanalysis. Ion microanalysis is based on the use of ion–ion emission, which consists of an ion beam's sputtering atoms from a target, some of which will leave as positive or negative ions. As large a fraction as possible of these ions is directed into a mass spectrometer, which transmits particles with a given e/M, i.e., ions of one of the elements in the target. By repeating this procedure with the mass spectrometer tuned to transmit particles with different values of e/M, one obtains information on the relative amount of different elements in the microscopic volume of the target under study.

A diagram of a modern microanalyzer of this type is shown in Fig. 9.6 [289]. Positive or negative ions formed in a hollow-cathode duoplasmatron are accelerated to energies of 5-22.5 keV. As it passes through the mass separator the ion beam is cleared of contaminant ions and neutrals. It then passes through alignment plates into the focusing structure. Because of the relatively large mass of the ions, at present ion beams can only be shaped with electrostatic lenses. The first einzel lens serves as a condenser, and the second is a short-focus objective lens which is used directly to produce an ion probe with a current i_p and radius r_p on the test target. Deflection plates are mounted in front of the second lens so that the beam can be used to scan selected parts of the target surface. The surface can also be observed with a built-in optical microscope. Another lens is installed to increase the number of secondary ions that enter the double-focus mass spectrometer. Positive or negative ions with a fixed e/M ratio are transmitted in the form of a beam with a current of

$$i = i_p Y \alpha^{\pm} \xi C_M, \qquad (9.3)$$

where Y is the number of sputtered atoms per primary ion, $\alpha^{\pm}$ is the degree of ionization of the sputtered atoms of a given element, ξ is the transport coefficient, equal to the ratio of the number of ions of a given element leaving the mass spectrometer to the number leaving the target surface, and C_M is the relative concentration of the element with atomic mass M. In this apparatus the secondary ions are not detected directly after the mass analyzer. They are further accelerated and then strike a converter electrode, knocking electrons out of it. These electrons are then accelerated and directed onto a scintillator with a photomultiplier. The output of the latter is a signal that is proportional to the current of secondary ions with a given e/M. The signals obtained in this way can produce an image on a cathode ray tube of the target in "lines of a given element" when the ion probe and display beam are scanned synchronously, they can be recorded on signal-time axes, or they can be read off a counter.

Mass spectroscopic analyses can, of course, be done without an ion probe by vaporizing and ionizing atoms of the test material by other methods. However, the advantage of an ion probe, whose localization and size can be controlled, is that it is possible to study the sample in three dimensions with high resolution, rather than globally. The best transverse resolution of an ion probe is determined by the diameter of the region occupied by the collision cascade produced by a single primary ion in a plane to a depth on the order of 1 nm, from which secondary particles can still escape. Under typical conditions this quantity, the diameter of an ideal ion probe, is roughly 10 nm.

Because of the need for a high ion current density $j_p = i_p/(\pi r_p^2)$, the displacement velocity of the target as it is sputtered, $dz/dt = j_p Y/en_a$, where n_a is the number of atoms/cm^3 in the matrix, is also important. Thus, for $j_p = 1$ A/cm^2, $Y = 1$, and $n_a = 5 \cdot 10^{22}$ cm^{-3}, we have $dz/dt \approx 10^3$ nm/s. Unlike electron probe analysis, therefore, ion-beam analysis is a "destructive" technique. In fact, this shortcoming is an advantage when a layer-by-layer analysis of a solid is needed. The depth resolution of an ion probe is related to the following factors:

(1) because of the nonuniform distribution of the current density over the beam cross section, the bottom of the crater formed by the beam is not flat;

(2) because of nonuniform erosion of the sputtered surface, it becomes rough; and

(3) diffusion, enhanced by ion bombardment, actually causes a distortion in the depth distribution of the target constituents.

In order to avoid the first effect, a thin scanning probe is used or only the information obtained from the flat part of the crater is used.

The minimum detectable relative concentration of a given element is given by

$$(C_M)_{min} = i_{min}/(i_p \alpha^{\pm} \xi Y), \tag{9.4}$$

where i_{min} is the minimum measurable ion current. In order to raise the sensitivity of this method the beam current, $\alpha^{\pm}$, and ξ must be increased as much as possible. For pure metals $\alpha^{\pm}$ is very small, but it can apparently be raised by several orders of magnitude by introducing high concentrations of electronegative (during analysis of positive ions) or electropositive elements (during analysis of negative ions). In particular, α^+ can be raised by exposing the target to oxygen or by using an oxygen ion beam as a probe. Coating the target with cesium makes α^- larger. (It is this fact which led to the development of intense surface-plasma sources of negative hydrogen ions.) Therefore, while the characteristics of the x-ray emission used in electron microanalysis are uniquely determined by electronic transitions among the inner shells of the atoms in the test sample, the secondary ion emission which carries the information in ion microanalysis is related to the physical and chemical properties of the nearest surrounding atoms. This circumstance makes it difficult to interpret ion microanalysis data since in order to determine C_M it is necessary to determine $\alpha^{\pm}$ in an independent experiment under identical conditions or to calculate it using certain model assumptions. The difficulty with ion microanalysis, therefore, arises in the dependence of secondary emission on the properties of the surrounding atoms, but it is just this factor which makes it possible to obtain information on the physical and chemical properties of a surface with an ion probe. The possibilities of obtaining such information are extended when positive and negative ions are used as probe and test particles. If we take $i_{min} = 10^{-18}$ A, $i_p = 10^{-8}$ A, $Y = 1$, and $\alpha^{\pm}\xi = 10^{-3}$ in Eq. (9.4), then $(C_M)_{min}$ is on the order of 10^{-7}. The minimum relative concentration and the minimum amount of a substance that can be measured with an ion probe are smaller than the typical corresponding values for electron probe analysis.

One of the most important features of ion microanalysis is that it can be used to the same extent for determining the amounts of both light and heavy elements, while it is difficult to measure light elements by electron probe analysis because the characteristic lines are so weak. Ion probe analysis also makes it possible to measure another important quantity, the isotopic ratio of elements in the sample.

Formation of ion probes. The basic components of the simplest system for forming an ion probe are an ion source and a short-focus lens. The minimum diameter of a low-current ion beam focused with aberration-free optics is limited, not by space charge, but by the thermal spread of the ion velocities and is equal to

$$d_{min} = \frac{d_0}{\sqrt{\mu}} \left(\frac{\chi\, T_i}{eU} \right)^{1/2} \frac{1}{\sin \theta}, \tag{9.5}$$

where d_0 is the initial beam diameter, T_i is the effective ion temperature, eU is the ion energy, μ is the ratio of the actual current density at the focus to the Langmuir limit current density ($\mu < 1$); and θ is the half angle of the beam cone in the focal region. It follows from Eq. (9.5) that in order to form a thin probe, d_0 (the diameter of the source aperture) and T_i must be as small as possible. In forming micron and submicron probes it is also impossible to neglect aberrations. The real beam diameter is

$$d = (d_g^2 + d_s^2)^{1/2}, \tag{9.6}$$

where d_g is the limit on the beam diameter owing to chromatic aberration and $d_s = C_s\theta^3/2$ is the spherical aberration (C_s is the spherical aberration coefficient). In order to reduce d_g the lens is moved further from the source, and in order to reduce d_s the input aperture of the lens is stopped. All this, however, may lead to an unacceptable reduction in the probe current. It can be kept large enough only when the source has a high brightness, i.e., a high current density per unit solid angle.

For focal spot sizes of less than about 10 μm the diameter of the spot is limited by chromatic aberration to [305]

$$d_g \sim C_g \, \alpha \Delta E / E, \tag{9.7}$$

where C_g is the chromatic aberration coefficient of the lens, α is the divergence angle, ΔE is the scatter in the ion energy, and E is the ion energy.

The current and current density (A/cm^2) in the focal spot are given by

$$I = \pi \alpha^2 \, dI / d\Omega \tag{9.8}$$

and

$$j \sim \frac{dI/d\Omega}{\Delta E^2} \left(\frac{E}{C_g} \right)^2. \tag{9.9}$$

It is important to note that the current density is independent of the input angle of the lens. A relatively large current density can be obtained by raising the ion energy, but it is limited by the electrical breakdown strength of the accelerating gap. Thus, the current density usually reaches 1 A/cm^2 at a field of 10^5 V/cm. With a hydrogen ion source, current densities on the order of 100 A/cm^2 can be obtained.

The system sketched in Fig. 9.6 and most other analyzers use a duoplasmatron, which is favorably distinguished from other plasma ion sources by a high output ion current density (up to 100 A/cm^2) and a relatively low spread in the ion thermal velocity in the appropriate operating regimes. In the present case a modified source was chosen, a duoplasmatron with a hollow cathode, which has a long operating life and is able to produce a discharge in electronegative gases. The diameter d_0 in duoplasmatrons varies from 100 to several hundred microns. Data have been published on a duoplasmatron with an output aperture of about 10 μm [290].

The focal length of the main lens used to form the probe beam is chosen to be minimal. But there is a limit to this length which is related to the purpose of the apparatus. For example, gaps large enough for secondary ion collection must be left between the lens and sample in microanalyzers.

Despite the need to reduce the spherical aberration in lenses for micron probes, a simplification in their manufacture or in their power supply circuits can sometimes be achieved at the cost of increasing the aberrations.

In one mass spectrometer with a duoplasmatron source [172] the thermal velocity spread of the Ar$^+$ ions corresponded to 4 eV, the brightness was 200 A/cm^2·sr, the focal length of the objective was 3 cm, $C_s = 100$ cm, and the probe

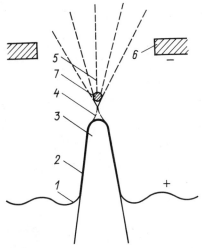

Fig. 9.7. A schematic illustration of a liquid-metal ion emitter: (1, 2) liquid metal; (3) metallic needle; (4) liquid-metal point formed in the electric field; (5) metal ions; (6) extractor; (7) visible emission region.

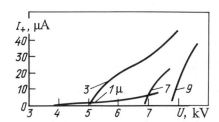

Fig. 9.8. Volt–ampere characteristics of a gold ion source with different needle radii.

diameter was less than 2 μm. With a design similar to that shown in Fig. 9.1, when the probe diameter was increased from 2 to 20 μm the current rose by a factor of 100 from $3.2 \cdot 10^{-9}$ A and the current density remained almost unchanged at 0.1 A/cm^2.

With a system that was not intended for microanalysis, a 1-μm-diameter beam with a current density of 0.5 A/cm^2 was obtained with a duoplasmatron and a single lens with a focal length of 10 mm. Two lenses were used to further reduce the beam diameter to 0.3 μm while keeping the current density at a level of 0.35 A/cm^2.

We note in conclusion that Eq. (9.4) implies that $C_M \sim 1/(j_p r_p{}^2) \sim 1/(\dot{z} r_p{}^2)$. This means, for example, that the lower the concentration of the emitting element is, the greater the amount of erosion of the sample that must be allowed during a measurement. An important conclusion can be drawn from this relationship to emphasize the need for developing high-brightness ion sources: When the probe radius is reduced by, say, a factor of 10, it is necessary to raise the current density of the probe by a factor of 100.

The development of ion sources has a complicated history, closely connected to the general development of gaseous-discharge physics, plasma physics, and surface physics. One of the most important stages in this history was the creation of a new type of emitter, the liquid-metal ion emitter, which has an especially high brightness.

Without going into technical details, the ion emitter can be regarded as a small part of a liquid-metal surface that wets a metallic needle and coats it with a thin film (Fig. 9.7). In front of the emitter there is an extractor electrode which creates a strong electric field (on the order of 10^8 V/cm) near the point to accelerate the ions. The extractor has a hole for removal of an ion beam. The intense emission from the metal ions observed under these conditions cannot be explained by any single known process, particularly field vaporization. In situations which can be treated in terms of the latter process, such as when gallium is on the surface of a tungsten needle in the solid phase or moves to the tip of the point from a reservoir of liquid gallium and does not accumulate there but creates an ion field emission image of the tungsten, the gallium emission current is less than 10^{-9} A. The liquid-metal ion emission regime is distinguished by high emission currents (10^{-6}-10^{-3} A), the existence on the needle surface of liquid metal which acquires a particular configuration in the high electric field, self-heating (up to 500-1000°C for gallium) in the emitting region, a characteristic luminosity near the needle, and an ion field emission image in the form of a massive spot or sometimes in the form of bright concentric rings which appear and disappear. Liquid-metal ion emission can be obtained by raising the temperature of the metal to the melting point T_m through external heating with a fixed, rather high negative potential U on the extractor or by raising U to a threshold value U_{thr} with $T > T_m$ [293] (Fig. 9.8). In both cases the resulting regime is always accompanied by a jump of several orders of magnitude in the ion current. After a threshold jump, the volt–ampere characteristic of an emitter (Fig. 9.8) has an average slope of $dI_+/dU \cong 10$-50 μA/kV.

The composition of beams extracted from liquid-metal emitters is of some interest [294, 295]. The most widely used gallium emitters produce mainly Ga^+ atomic ions (about 99%) and a small number of Ga^{2+} ions, while a gold emitter differs in having a large content of doubly charged ions (65% Au^+ and 20% Au^{2+}). As the overall emission current increases with changes in the accelerating potential, there is an increase in the number of emitted atomic ions as well as a rapidly increasing, relatively small component of molecular ions. The emitter produces neutral as well as charged particles. That neutrals exist near liquid-metal emitters is confirmed by the observed luminosity from neutral atoms and, in the case of a cesium emitter, by the signals from a detector based on surface ionization of cesium atoms. Liquid-metal emitters can also serve as sources of metal droplets.

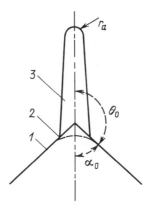

Fig. 9.9. The shape of liquid-metal emitters: (1) Taylor cone; (2) Taylor cone with a rounded vertex; (3) most probable shape of the emitter (a Taylor cone with a stream escaping the vertex).

Since liquid-metal emitters operate in a high electric field, one fairly sensitive method for determining the places or origin of the ions is to study their energy distribution. This is also a basic source of information on the mechanism for ion emission [294]. Measuring the energy deficit of an ion as it is slowed down makes it possible to determine the potential difference between the formation zone and the liquid metal and, therefore, to determine the location of this zone.

The energy deficit of an ion formed on an emitter by electric field vaporization is $\Delta E = I_n - n\varphi + U_0 - Q(E)$, where I_n is the ionization potential of the n-fold charged ion, φ is the the work function of the retarding electrode-collector, and U_0 is the evaporation energy. For Ga^+ with an activation energy $Q(E) \approx 0$ the calculated $\Delta E = 4.9$ eV. The energy distribution of the Ga^+ ions behaves as follows for different emitter temperatures: at low temperatures the measured energy deficit $\Delta E = 4 \pm 1$ eV corresponds to the appearance of an ion during evaporation by the electric field. As the emitter temperature is raised, the main peak in the distribution moves toward lower ΔE, presumably because of excitation of gallium atoms on the surface by electrons which are formed by ionization in the (volume) electric field and move toward the emitter. The appearance of an additional broad peak with an energy deficit of 10 eV is explained by electric field ionization of the increasing flux of neutral atoms. The observed growth in the half width of the ion energy distribution with an overall current I_+ [292] is caused by the conversion of transverse momentum into longitudinal by Coulomb collisions in the dense diverging ion beam. Despite an increase in the solid angle Ω, the angular intensity of the beam (I_+/Ω) increases with I_+ and amounts to 10-50 µA/sr [295]. The energy distribution of the molecular ions, especially those of gold, has a second peak with an energy as high as 300 eV, which is evidence that these ions are formed at a considerable distance from the needle [294].

In sum, experimental data on the ion energy distribution have been a source of valuable information. There are three zones in which ions are formed. At low currents atomic ions are mainly formed immediately at the emitter and at higher currents, on the emitter and partly near it at distances of 1-10 nm. Molecular ions are mainly formed at distances of 10-100 nm from the emitter. All of this suggests that ions are formed both during electric field vaporization on the needle surface and by electric field ionization at some distance from the surface.

Finally, note should be taken of the absence of reliable data on the transverse dimensions of the emission zone. This is yet another quantity which must be known in order to explain the physical nature of this phenomenon. Many authors adhere to the opinion of Gomer [301] in estimating the radius of this zone to be 10^{-7} cm, although it is difficult to bring this sort of estimate, without any experimental confirmation, into agreement with general concepts. The actual configuration of the emitter and the dimensions of the emission zone cannot be clarified without examining the stability of liquid-metal surfaces in strong electric fields.

Instability of liquid-metal surfaces in electric fields. The work of Taylor [296] is usually considered the basis for the assumption that at the critical potential difference U_{cr} between an extractor and a liquid-metal emitter the latter acquires a conical shape with a definite vertex angle $2a_0$ (Fig. 9.9). The equilibrium condition for the pressures on the surface of conducting liquid with surface tension σ and electric field E has the form

$$\sigma(1/\rho_1 + 1/\rho_2) = E^2/8\pi, \tag{9.10}$$

where ρ_1 and ρ_2 are the principal radii of curvature. Since the only radius of curvature of a conical surface is inversely proportional to the distance R from the vertex, Eq. (9.10) implies that in equilibrium the electric field falls off as $R^{-1/2}$. Equilibrium is realized in a field with the potential

$$U = CR^{1/2} P_{1/2} (\cos \theta), \tag{9.11}$$

where C is a constant and $P_{1/2}(\cos\theta)$ is the Legendre function. From the condition that the cone be an equipotential $P_{1/2}(\cos\theta) = 0$, we find $\theta = \theta_0 = 130.7°$ and $\alpha = \alpha_0 = 49.3°$. The equality $\sigma\cot\alpha_0/R = (8\pi R)^{-1}(\partial U/\partial\theta)$, which follows from Eq.(9.10), makes it possible to determine the field normal to the surface of the Taylor cone (V/cm),

$$E = \frac{1}{R} \frac{\partial U}{\partial \theta} = 1.4 \cdot 10^3 \; \sigma^{1/2} R^{-1/2}, \tag{9.12}$$

eliminate C, and calculate the critical potential difference necessary for forming such a cone (V),

$$U_{cr} = 1.4 \cdot 10^3 \; \sigma^{1/2} R_0^{1/2}. \tag{9.13}$$

Here R_0 is the value of R for $\theta = 0$, or the distance from the vertex of the cone to an extractor whose shape is given by $R = R_0[P_{1/2}(\cos \theta)]^{-2}$.

A variety of experiments confirm the transition from a spheroidal surface to a conical one at some critical field. Liquid-metal emitters which have hardened in the form of Taylor cones have actually been observed, although in a number of cases the vertex angle was substantially different from $2\alpha_0$. A 200-keV electron microscope has been used to study the shape of the needle tip directly during emission. A small conical protrusion with an angle of about 90° was observed. At the same time, it is difficult to match an emitter in the shape of an unbounded Taylor cone with the feasible emission mechanisms.

Calculations show that a field $E \approx 10^8$ V/cm is needed to obtain ions by field evaporation or field ionization. Equation (9.9) shows that this field is localized on a gallium ($\sigma = 7 \cdot 10^{-3}$ N/cm) Taylor cone in an emission zone with a radius on the order of 10^{-7} cm. This improbably small radius for the emission zone is specific to a conical emitter, for which the field falls off sharply with distance from the vertex. The idea of an emitter of this shape must be examined critically. Attempts at a numerical determination of the strength of the self-consistent electric field, including the ion space charge near an emitter in the form of a Taylor cone, were unsuccessful even with a minimum current $I_+ = 1$ µA. When space charge was taken into account, the field changed sign near the vertex, which excluded the possibility of current flow and, therefore, contradicted the initial data. At the same time, a model of an emitter in the form of a Taylor cone with a protrusion at its vertex that ends in a hemisphere (Fig. 9.8) yielded the desired self-consistent field distribution with space charge taken into account.

The need to eliminate these contradictions has stimulated interest in the stability of liquid-metal surfaces in electric fields. Tonks and Frenkel predicted the instability of gravitational-capillary waves on a liquid-metal surface in the presence of a sufficiently high electric field. The instability develops because a small perturbation of the liquid, $\xi = a \exp[i(kx - \omega t)]$, produces both the additional pressures $\rho g \xi$ and $\sigma \partial^2 \xi / \partial x^2$ and a negative pressure $E^2 \xi / (4\pi)$, which leads to a growth in the initial perturbation, to a new rise in the negative pressure, and so on. From the dispersion equation for gravitational-capillary waves in the presence of an electric field,

$$\omega^2 = \frac{k}{\rho}\left(\sigma k^2 + \rho g - \frac{E^2}{4\pi}k\right) \tag{9.14}$$

we obtain the well-known formula for the critical field of the aperiodic Tonks–Frenkel instability ($\omega^2 < 0$),

$$E_{cr}^{T-F} = (64\,\pi^2\,\sigma\rho g)^{1/4}. \tag{9.15}$$

For the conditions examined below, only the short-wavelength capillary waves ($\sigma k^2 > \rho g$), which are unstable at the critical field

$$E_{\text{cr}}^{\sigma} = (4\pi\sigma k)^{1/2}, \tag{9.16}$$

are of interest. This equation implies that liquid-metal emitters in their characteristic high electric fields are unstable, even relative to very short-wavelength perturbations. In his theoretical analysis Taylor did not touch on questions of instability, but he did demonstrate experimentally an instability of the cone vertex which manifested itself in the emission of a very thin stream of liquid. The idea that the actual emitter is not the smoothed vertex of a cone, but the front of a stream flowing out of the vertex, has been proposed by Krohn and Ringo [298]. This idea is in agreement with theoretical and experimental studies of the nonlinear stages of the Tonks–Frenkel instability.

Summarizing the theoretical work on this point, we can predict two stages in the development of the instability. The first consists in the formation of a spheroidal or conical protrusion on the liquid-metal surface. In the second stage the volume of the protrusion is maintained and no additional liquid enters it, but the concentration of the electric field causes rapid growth of a peak on the protrusion with a reduced radius of curvature. Near the leading front it is possible to observe the development of a constriction which is the precursor of the breakoff of a droplet from the growing peak. This pattern is reproduced experimentally. The growth of microscopic protrusions on the surface of a liquid-metal in a strong electric field has been examined by shadow photography (high-speed camera) [299]. For all the attractiveness of this method, which makes it possible to directly follow the development of the instability *in situ*, it has certain limitations, as do all optical methods, and cannot provide information about very small details of the developing peaks with characteristic dimensions on the order of the wavelength of the light. In this sense the "freezing" of capillary waves for later study with a high-resolution scanning electron microscope is preferable [287]. A flux of hydrogen plasma propagating toward a copper slab with a negative potential fulfilled two functions simultaneously: the ions collected from the dense plasma heated the opposite side of the slab to the melting point and a strong electric field developed in the thin screening layer above the molten copper surface. Under the action of the net uncompensated momentum transferred to the surface by the field and particles, the molten metal was thrown onto a "cold beach" (consisting of the unmelted part of the slab) where it rapidly solidified. In this way it was possible to record the excitation of nonlinear capillary waves and study them later with the microscope. A panorama of such frozen waves and isolated fragments of them is shown in Fig. 9.4. This confirms the formation of spheroidal or conical protrusions on which peaks with reduced radii of curvature subsequently developed. Droplets that were frozen at the moment of breakaway are visible on several peaks. The shape of the layer of metal that coats the emitter needle is directly related to this instability.

It is, therefore, extremely likely that ions are collected from the front of a very fine stream of metal emitted from the vertex of a cone as a result of this instability, rather than from the smoothed vertex of a Taylor cone (see Fig. 9.9). In this case the field on the emitter can be calculated using a formula from field emission microscopy, $E = U/\mu r_a$, where μ is a coefficient roughly equal to 5 over a wide range of values of R_0/r_a (from 10^3 to 10^5), and r_a is the radius of curvature of the stream front. Since the field is relatively uniform under these conditions, the value $r_a =$

10^{-5} cm calculated for typical conditions ($U = 5$ kV, $E = 10^8$ V/cm) must be close to the radius r of the emission zone.

The current density $j_+ = 10^6$ A/cm^2 corresponding to a radius of 10^{-5} cm (for $I_+ = 100$ μA) does not provoke the objections based on the effect of space charge that were raised above. Of course, for such a relatively large value of r_a and a fixed value of E, the equality $E^2/(8\pi) = 2\sigma/r_a$ may not be satisfied, i.e., the pressure on the surface may be unbalanced. It must be concluded that the concept of a sharp-tipped emitter unavoidably implies that the configuration and location of the emitting boundary of a liquid-metal ion source is determined solely by the equality of the number of atoms incident on the emitter to the number of atoms leaving it, and not by the pressure balance equation (9.7) [300].

If we assume that there is a threshold potential at which an instability develops on the metal surface with a maximum wavelength equal to the radius of curvature of the needle, then by setting $E = f(r)$ in Eq. (9.4), we obtain an expression for this potential,

$$U_{\text{thr}} = (8\pi^2 \mu^2 \sigma r_n)^{1/2}, \tag{9.17}$$

which is in qualitative agreement with experimental data [293] for "sharp" needles (see Fig. 9.8). As can be seen by comparing Eqs. (9.13) and (9.17), the threshold potential is not necessarily the same as the critical potential at which a Taylor cone is formed.

The above information suggests that the ion flux generated by a liquid-metal emitter is nonuniform in its origin. The flux is primarily ions formed during evaporation by the electric field. As the total emission current (accelerating potential) increases, ionization by the electric field becomes noticeable; ions are formed in the volume near the emitter. The resulting electrons are accelerated by the electric field and, on striking the emitter, heat it to the observed high temperature. Molecular ions probably are produced by a charge exchange process [294]: $\text{Me}^+ + \text{Me}_n = \text{Me}_{n+1}^+$ or $\text{Me}^+ + \text{Me}_n = \text{Me}_n^+ + \text{Me}$. Me_n clusters are formed both on the liquid-metal surface and in the flux of moving neutral particles. When the current is raised, the formation of droplets with diameters as large as several microns becomes more probable.

Limit on the ion current density and beam current. With this general picture of the ion production process in mind, it is possible to examine the important question of limitations on the ion current density and the total beam current. We have already pointed out the two possible steady states of an emitter. According to one widespread point of view, the shape and location of the emitter is determined by the condition (9.10). However, another state besides this one is possible. There the pressure balance is not obeyed, but $E^2/(8\pi) > 2\sigma/r_a$ and the location of the emitter is determined solely by the dynamic equilibrium of the particle flux at its boundary with the vacuum [300], i.e., the equality between the flux $\rho v/M$ of atoms reaching the surface and the flux j_+/e of emitted ions. The latter condition makes it possible to calculate the maximum current density of the ions produced by electric field evaporation. When the metal atoms arriving at the surface are completely converted into free ions, the conservation of momentum implies that

$$\rho v v_1 \leqslant E^2/(8\pi), \tag{9.18}$$

where v_1 is the average velocity of the emitted ions and v is the velocity of the metal. This inequality imposes a limit on the ion current density:

$$j_+ = e\rho v/M \leqslant eE^2/(8\pi M v_1). \tag{9.19}$$

The typical parameters $E = 10^8$ V/cm, $v_1 = 4 \cdot 10^5$ cm/s, and $M = 10^{-22}$ g correspond to a maximum ion current density of $j_{+,max} = 10^7$ A/cm.

If, therefore, we assume that the ions are emitted by the equilibrium surface of a Taylor cone acted on by the field of Eq. (9.12), then the calculated ion emission current will be smaller than the observed value: $I_+ = j_{+,max}\pi r_p^2 < 10^{-7}$ A. If, on the other hand, we assume that the emitter has a pressure imbalance and an emitting zone with a radius of 10^{-5} rather than 10^{-7} cm, then the calculated ion current agrees with experiment: $I_+ = j_{+,max}\pi r_p^2 < 10^{-3}$ A. This is yet another argument in favor of the concept of a sharp-tipped emitter with a pressure imbalance on its surface.

Although liquid-metal emitters are located in a strong electric field, because of the high current density the ion space charge is substantial even for the minimum current. By reducing the electric field strength, the space charge creates a negative feedback in terms of which Gomer [301] explains the high stability of the ion current collected from liquid-metal emitters. The space charge can indeed damp the current oscillations, but a more convincing explanation [302] is that the noise in the ion beam is limited by fluctuations in the flux of metal to the emitter.

Space charge also determines the maximum current in an ion beam whose source is a point with a given potential difference U_0, between it and the extractor. In order to calculate the field including the space charge, the simplest model of two concentric spheres, an inner emitter with radius r_a and an outer extractor ($R_0/r_a = X$), is invoked [301, 303]. In terms of dimensionless parameters (the potential $y = U/U_0$, and radius $x = r/r_a$) the Poisson equation can be written in the form

$$\frac{d^2y}{dx^2} + \frac{2}{x}\frac{dy}{dx} = \frac{A}{x^2(y+\alpha)} \tag{9.20}$$

with the boundary conditions $y(1) = 0$, $y(x) = 1$. Here $A = I_+(M/2e)^{1/2} U_0^{-3/2}$ and $\alpha = Mv_1^2/(2eU_0)$, where e and M are the ion charge and mass and v_1 is their initial velocity. The dependence of the field at the emitter $dy/dx|_{x=1}$ on A is found numerically on a computer and can also be approximated by substituting the solution of the appropriate Laplace equation for the function y on the right-hand side of Eq. (9.17). From the approximate solution

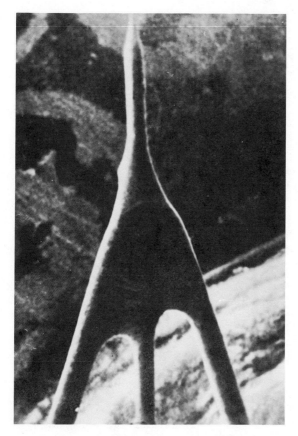

Fig. 9.10. A photograph of a liquid-metal gallium ion emitter.

$$\left.\frac{dy}{dx}\right|_{x=1} \cong [1 - A(\ln X - 1)] \tag{9.21}$$

it follows, first of all, that in this electrode system the drop in the space-charge field depends weakly on the radius of curvature of the emitter (X). This justifies treating the problem with a fixed radius r_a. The evident limitation on A makes it possible to estimate the maximum ion current which can be collected from a single point in a real spherically symmetric system. Thus, for $U_0 = 6$ kV and $M = 10^{-22}$ g, the calculated maximum current is less than 1 mA, in agreement with experiment.

The volt–ampere characteristic of a liquid-metal emitter is determined in a complicated fashion by the space-charge limitation of the field, the dependence of the emission current on the field, the interdependence of the shape of the point and the electric field strength on the surface of the point, and the impedance to the flow of the viscous liquid along the base of the emitter, the side surface of the needle [293].

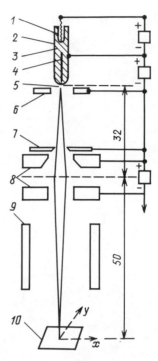

Fig. 9.11. A system for producing a sub-micron gallium ion beam: (1) heater; (2) reservoir; (3) liquid gallium; (4) tungsten needle; (5) emitting point; (6) extractor; (7) aperture; (8) electrodes of accelerating lens; (9) deflecting plates; (10) target.

An exact solution of Eq. (9.20) also reveals another interesting feature of the space-charge behavior: at relatively high currents the maximum of the electric field is not on the emitter surface, but at some distance from it.

Liquid-metal emitters. Liquid-metal ion sources can be divided into three groups in terms of their structural features. In the first type of source liquid metal is fed through a capillary [298], at whose outlet an ion-emitting surface of this metal is formed by an external electric field. The second type of source is distinguished by the fact that liquid metal is kept in a special reservoir, from which a metal needle (usually tungsten) protrudes. The surface of the needle is wetted and coated by a thin layer of the liquid metal which is the actual ion emitter. The third group of sources includes those with a metal needle welded onto a metal loop (Fig. 9.10). The temperature of the needle is controlled by passing a current through the loop. At the same time, the needle and loop serve as a reservoir of liquid metal which covers their surfaces more or less uniformly, although it occurs in one location as a droplet. As in the second type of source, the emitter is the surface of the liquid metal that wets the tip of the needle in a thin layer. In each device, an extractor is placed in front of the needle, and sometimes another electrode, a control grid, is placed in front of the extractor. Prior to operation of the source the needle surface is cleaned until it is thoroughly wetted by the liquid metal [293]. The needle material must, of course, not dissolve in the liquid metal. Particular difficulties arose in the development of aluminum ion sources because of the active interaction of molten aluminum with most metals, including tungsten. In this case a titanium-worked graphite needle was successfully used.

In the preceding discussion we have presented data on gallium and gold emitters as typical. However, studies have been made of liquid-metal emitters for ions of Ga, In, Au, Cs, Al, Bi, Zn, Sn, Hg, Ge, U, Pt, Fe, and other metals. When the obstacle to making an emitter is the high melting point of the metal or its high vapor pressure, alloys with melting points below those of the initial constituents, such as Bi–Sn–Cd, Au–Ge, Sb–Au–Pb, and Au–Si, are used.

The extremely high current density of liquid-metal ion emitters (10^4-10^6 A/cm^2) yields a brightness of $\approx 10^6$ A/(cm^2·sr), or four orders of magnitude higher than that of the best classical plasma sources such as the duoplasmatron. The relatively large scatter in the ion energies obtained from these emitters can lead to substantial chromatic aberration. In this regard, as well in connection with droplet formation, the correct choice of ion-beam current (accelerating potential) is important.

Figure 9.11 shows an example of a scheme for producing a submicron ion probe using a liquid-metal gallium ion source [304]. This kind of system can be used to form an ion probe with a diameter of 10^{-5} cm, a current density of 1.5 A/cm^2, an energy of 57 keV, and a brightness at the target of $3.3 \cdot 10^6$ A/(cm^2·sr). In a system for studying the condensation of gallium ions carried by an ion beam, a liquid-metal emitter was employed only for low currents (on the order of microamperes), when the formation of clusters or droplets, which were unacceptable in this case, could be avoided with certainty. Recently a three-electrode system has been used to form ion beams with diameters of 0.1-0.5 µm and energies that could be varied over a wide range (20-2 keV).

9.3. Prospects for the Use of Submicron-Sized Ion Beams

Possible applications of ion microprobes. Liquid-metal emitters can be used in widely different areas of science and technology: in scanning ion spectroscopy; in secondary ion–ion emission mass spectroscopy; for vaporization of x-ray and optical masks; in ion-beam lithography; for implantation of dopant elements in semiconductors without the use of masks; for adjustment of critical currents in Josephson junctions; for polishing, layer analysis, and dimensional micromachining; for making thin films and small grain powders; in charged-particle accelerators; and in thermonuclear experiments.

Scanning ion microscopy. A focused ion beam can be used to make an image of a sample as with a scanning electron microscope. Low-energy secondary electrons are knocked out of the surface by an ion beam. They are detected by channel electron multipliers or by a scintillator and photomultiplier combination. An image of the surface is obtained by scanning it with the ion beam. The high current density and submicron size of the beam provide high resolution. At Bell Laboratories (USA) a 20-keV In$^+$ ion beam has been used on a scan field of 173 µm to obtain an image of a stainless steel grid with a step size of 200 lines/inch placed on a copper grid with a cell size of 1000 lines/inch. The resolution was 1 µm. A silver film deposited by vapor deposition through the grid was clearly resolved on a silicon grid, which illustrates the chemical contrast [305].

Correcting masks. Integrated circuits have become large and extremely complicated, so it is very important to have a means for correcting defects in lithographic masks. Defects in chromium optical masks are corrected by laser beam vaporization (flash off). Lasers have a resolution limit of about 1 µm for correcting masks with opaque defects. The density of defects in x-ray masks is an order of magnitude greater than in optical masks. Laser correction of these defects is made difficult by large thermal diffusion. One new method for correcting mask defects is to use a sharply focused ion beam which produces an image of these defects and simultaneously removes them by sputtering. This method has resolved patterns with a width and space between lines of 0.25-0.5 µm on chromium films. The rate of removal of the defects was 1 µm^2 in 10 s. Sputtering is not a thermal process, so that the resolution is independent of the thermal conductivity. Finally, an ion beam can remove both dark and light defects.

Ion-beam lithography. Ion microprobes can draw masks with the high resolution of the primary patterns. Mask patterns can be transferred onto a substrate using channeled ions (contact prints) or ion-optical imaging (reduced prints). As opposed to electron lithography, ions are scattered less in the resist material and create very low energy secondary electrons, so that the resolution is improved. Special resists are required for electron-beam lithography, while for ion-beam lithography the requirements for the resist are the same as for ordinary polymers. For example, a pure novolac resin may be exposed by an ion beam. It is more sensitive to ions than to electrons, since the ions have a high atomic number and considerably lower velocity. One disadvantage of ion-beam lithography is the long time for inscribing the image, from 2 min to 1 day for images with dimensions of 0.5-0.1 µm at ion current densities of 1 A/cm^2.

Assuming that a quality write requires 1000 ions on a point and a form density of 50%, we find the minimum inscription time to be (s/cm^2)

$$t = 4/(F^4 j),\qquad\qquad(9.22)$$

where F is the minimum dimension of the inscription (µm) and j is the current density (A/cm^2).

The use of silicon dioxide films as a resist for ion lithography has been studied by Valiev et al. [306]. Its physical and chemical properties change significantly during ion bombardment: the refractive index increases, the rate of chemical etching rises, and the density and vibrational spectrum change. For example, during bombardment of 30-nm-thick silicon dioxide films with 50-keV nitrogen ions at doses of 10^{13}-10^{15} cm^{-2}, the selectivity of etching in a water solution of hydrofluoric acid was determined from the ratio of the rates of etching of irradiated and unirradiated surfaces and changed from 5.5 to 1.2. The selectivity of etching and changes in the vibrational spectrum were slowed down and terminated at doses on the order of $5 \cdot 10^{14}$ cm^{-2} and remained constant up to doses of 10^{16} cm^{-2}. The average number of radiation defects produced in a single collision cascade is given by

$$\overline{N_g} = 0.4 \, E_0 / (2E_d),$$ (9.23)

where E_d is the average energy, 10 eV, below which a particle no longer contributes to the cascade. Under these conditions N_g is 1000 defects per cascade. The number of defects in an irradiated volume at which "saturation" of etching selectivity sets in is given by

$$N_{max} = \overline{N_g} \, D_{sat} \, F,$$ (9.24)

where D_{sat} is the saturation dose and F is the irradiated area. The volume of the irradiated region of the film is $V = FR_p$, where $R_p = 1.1 \cdot 10^{-5}$ cm is the mean projected path. The recombination volume of a single defect is given by $V/N_{max} = R_p/(N_g D_{sat}) \approx 2 \cdot 10^{-23}$ cm^3.

Studies have also been made of the etching of polymer resist films during bombardment by 25-200 keV ^{14}N$^+$ ions at doses of 10^{13}-10^{16} cm^{-2} [307]. Films of polymethyl methacrylate (PMMA) and polyhexenesulfone (PHS) with thicknesses of 0.4-1.3 μm were also used. The thickness of the PMMA layer removed by etching is given by

$$\Delta d = B \log (D/D^*),$$ (9.25)

where B is the thickness of the layer removed by a dose $D = 10D^*$, and D^* is the dose determined by extrapolation of the linear portion of the experimental dependence $\Delta d = f(D)$ to the abscissa. It is clear that the reduction in thickness of the resist during ion bombardment is caused by the appearance of volatile fragments in the volume of the resist as a result of radiation chemical transformations and diffusion of these fragments to the surface where they leave. On the other hand, as the volatile fragments diffuse to the surface they may undergo partial recombination (cross linking with the polymer chain).

Differentiating Eq. (9.25) with respect to D, we obtain the etching rate for the resist,

$$\nu_{etch} = 0.43 \, B/D.$$ (9.26)

The reciprocal dependence of the etching rate on the irradiation dose is explained by the dominance of cross linking of the molecular fragments as the dose is raised and by a reduction in the rates of destruction processes. When the film thickness d_0 is less than the projected ion path R_p, the relative change in the thickness of the layer removed by etching, $\Delta d/d_0$, reaches 0.7-0.8 in the ordinary resists that were studied. The remaining film is "graphitized" (or polymerized), and its hardness and chemical durability increase. Because etching is brought to a halt during irradiation by medium-energy nitrogen ions when the irradiated layer has reached some residual thickness, a combination of PMMA and PHS cannot be used for "dry" lithography.

The possibility of creating resists for which the destruction processes will become dominant at higher doses is not excluded. These resists would be convenient for direct ion lithography based on etching. Masks of these resists obtained by electron, x-ray, ion, or vacuum ultraviolet lithography could be used for local ion implantation in semiconductors. Corrections must be introduced in the dimensions of the topological elements for ion doping through these masks, however, since the dimensions will change during ion bombardment. For example, the period of a lattice in a PMMA layer was 2.24 ± 0.02 µm. After irradiation of part of a test lattice by 50-keV nitrogen ions to a dose of $5 \cdot 10^{14}$ cm^{-2}, the groove width changed from 0.3 to 0.6 µm.

Accelerators and thermonuclear experiments. Intense ion beams from liquid-metal emitters are also used in accelerator technology and in controlled thermonuclear fusion research [308-310]. One way of obtaining intense ion beams is to use an array of liquid-metal points whose surface serves as a source of dense plasma during a discharge pulse.

The maximum current density during evaporation or desorption by an electric field in the space-charge-limited regime for a single point is given by

$$j_{max} \sim \frac{4}{9 \pi r_0^2} \left(\frac{2e\,Z}{m} \right)^{1/2} \left(\frac{U}{k} \right)^{3/2}. \tag{9.27}$$

In a pulsed regime with $U \sim 10^7$ V, $E_0 \approx 7 \cdot 10^8$ V/cm, $r_0 = 5 \cdot 10^{-3}$ cm, and $k = 3$, $j_{max} \sim 10^7 \sqrt{Z/A}$, where Z and A are the charge and atomic mass of the ion. The maximum current under these conditions is

$$I_{max} \sim \pi r_0^2 j_{max} \text{ or } I_{max} \sim 10^3 \sqrt{ZA}. \tag{9.28}$$

In order to obtain an ion current of 10^5-10^6 A, 10^3-10^4 points must be used.

When the distance between the point anode and the plane cathode $D \gg r_0$, the maximum current is determined by the "three halves" law for a plane diode. For $U = 10^7$ V, $D = 3$ cm, and a plane cathode with an area on the order of 10^4 cm^2, $I \sim 10^6 \sqrt{ZA}$.

The formulas derived above are convenient for field desorption of elements that have been specially deposited on the surface of points (e.g., D_2, Li). The total current of desorbed ions (A) can be estimated as

$$I_g \sim 10^{-4}\, n Z\, \pi r_0^2\, N \tau^{-1}, \tag{9.29}$$

where n is the number of deposited atomic layers and τ is the duration of desorption. For $n \sim 10^2$, $\tau \sim 10^{-8}$ s, $r_0 = 5 \cdot 10^{-3}$ cm, and $N \sim 10^4$ we have $I_g \sim 10^6 Z$. Ion beams with currents of 10^5-10^6 A are needed for controlled thermonuclear fusion research and for making super-high-current charged-particle accelerators.

REFERENCES

1. M. D. Gabovich, *The Physics and Engineering of Plasma Ion Sources* [in Russian], Atomizdat, Moscow (1972).
2. N. N. Semashko et al., *Injectors of Fast Hydrogen Atoms* [in Russian], Énergoizdat, Moscow (1981).
3. I. M. Kapchinskii, *Particle Dynamics in Linear Resonance Accelerators* [in Russian], Atomizdat, Moscow (1966).
4. M. Dembinski, P. K. John, and A. G. Ponomarenko, Intense ion beam extracting from a flowing boundary, in: *Proc. 3rd International Conf. on High Power Electron and Ion Beam Research*, Novosibirsk (1979), Vol. 1, pp. 277–290.
5. B. N. Makov, Formation of a stationary electric field in the magnetized column of an ion source discharge, in: *Abstracts of Talks at 5th All-Union Conf. on Plasma Accelerators and Injectors*, Nauka, Moscow (1982), pp. 82–83.
6. F. M. Bacon, Gas discharge ion source. I. Duoplasmatron, *Rev. Sci. Instrum.*, **49**, 427–434 (1978).
7. N. I. Bortnichuk et al., Modeling of electron-optical systems with a plasma emitter, *Zh. Tekh. Fiz.*, **47**, 1894–1903 (1977).
8. *Proc. 2nd Symp. on Ion Sources and Formation of Ion Beams*, Berkeley, USA (1974).
9. M. D. Gabovich, Ion beam plasmas and the propagation of intense compensated ion beams, *Usp. Fiz. Nauk,* **121**, 259–284 (1977).
10. M. D. Gabovich, Compensated ion beams, *Ukr. Fiz. Zh.*, **24**, 257–273 (1979).
11. V. S. Anastasevich, Theory of compensated space charge in steady-state ion beams, *Dokl. Akad. Nauk SSSR*, **105**, 442–444 (1955).
12. M. D. Gabovich, Spreading of intense quasineutral positive ion beams, *Fiz. Plazmy*, **2**, 163–165 (1976).
13. M. D. Gabovich, L. P. Katsubo, and I. A. Soloshenko, Optimum focusing of compensated positive ion beams, *Fiz. Plazmy*, **3**, 614–618 (1977).
14. M. D. Gabovich, Limiting current density in intense focused quasineutral ion beams, *Zh. Tekh. Fiz.*, **46**, 1731–1736 (1976).

15. G. A. Cottrell, Magnetic effects in high current ion beams, *Rev. Sci. Instrum.*, **52**, 1174–1181 (1981).

16. M. D. Gabovich et al., Possible method for neutralizing space charge in high-power electron beams in ultrahigh vacuum, *Pis'ma Zh. Tekh. Fiz.*, **5**, 581–583 (1979).

17. M. D. Gabovich and A. P. Naida, Potential on an isolated body bombarded by a beam of positive or negative ions and some features of using the latter, *Ukr. Fiz. Zh.*, **15**, 864–866 (1971).

18. V. F. Virko et al., Synthesized plasmas obtained by neutralizing an ion beam with electrons injected through the ion source, *Zh. Tekh. Fiz.*, **44**, 2296–2301 (1974).

19. M. D. Gabovich and A. P. Naida, Collective oscillations in synthesized plasmas consisting of beams of positive and negative ions, *Zh. Eksp. Teor. Fiz.*, **60**, 965–971 (1971).

19a. M. D. Gabovich and A. P. Naida, Nonlinear effects in synthesized plasmas consisting of beams of positive and negative ions, *Zh. Eksp. Teor. Fiz.*, **62**, 183–188 (1972).

20. V. S. Imshennik et al., Mathematical simulation and experimental analysis of nonlinear interaction of positive and negative ion beams, *Comp. Meth. Appl. Mech. Eng.*, **9**, 1–23 (1976).

21. M. D. Gabovich, A. P. Naida, and É. A. Pashitskii, Injection of neutralized beams consisting of positive and negative ions into a filled plasma mirror, *Ukr. Fiz. Zh.*, **18**, 1744–1745 (1973).

22. M. D. Gabovich et al., Combined electron and ion beams. A model for the evolution of collisionless plasmas, *Fiz. Plazmy*, **8**, 808–810 (1982).

23. L.S. Simonenko, I. A. Soloshenko, and I. A. Shkorina, Excitation of electron oscillations in plasmas by a fast negative ion beam, *Ukr. Fiz. Zh.*, No. 11, 1886–1891 (1974).

24. M. D. Gabovich, I. A. Soloshenko, and A. F. Tarasenko, Effect of plasma oscillations in the electron background on the degree of compensation of ion beams, *Fiz. Plazmy*, **8**, 282–286 (1982).

25. A. V. Mikhailovskii, *The Theory of Plasma Instabilities* [in Russian], Atomizdat, Moscow (1970).

26. M. D. Gabovich, I. A. Soloshenko, and A. A. Goncharov, Ion plasma oscillations in "ion-beam" plasmas, *Zh. Tekh. Fiz.*, **43**, 2292–2296 (1973).

27. M. D. Gabovich et al., Excitation of ion plasma oscillations by a negative ion beam, *Zh. Eksp. Teor. Fiz.*, **67**, 1710–1716 (1974).

28. M. D. Gabovich, L. P. Katsubo, and I. A. Soloshenko, Dispersion of ion oscillations in "ion-beam" plasmas and measurement of the cold ion density, *Zh. Tekh. Fiz.*, **44**, 2286–2289 (1974).

29. M. D. Gabovich, D. G. Dzhabbarov, and A. P. Naida, Decompensation of dense negative ion beams, *Pis'ma Zh. Eksp. Teor. Fiz.*, **29**, 536–539 (1979).

30. L. P. Katsubo, V. P. Kovalenko, and I. A. Soloshenko, Spatial and temporal focusing of ion beams during excitation of transverse ion plasma oscillations, *Zh. Eksp. Teor. Fiz.*, **67**, 110–117 (1974).

31. V.I. Pistunovich et al., Charge exchange instability of dense ion fluxes in gases, *Fiz. Plazmy*, **2**, 3–23; 750–755 (1976).

32. M. D. Gabovich et al., Dynamic decompensation of positive ion beams in the absence of a magnetic field, *Fiz. Plazmy*, **6**, 925–932 (1980).

33. M. V. Nezlin, Plasma instabilities and the compensation of space charge in an ion beam, *Plasma Phys.*, **10**, 337–358 (1968).

34. B. E. Gavrilov, A. V. Zharinov, and V. I. Raiko, Dynamic decompensation of ion-beam space charge in electromagnetic isotope separation [in Russian], Moscow, Preprint No. IAE–592 (1964).

35. M. D. Gabovich, L. P. Katsubo, and I. A. Soloshenko, Gas compensation of positive ion bursts in the absence of a magnetic field, *Fiz. Plazmy*, **4**, 1370–1376 (1978).

36. M. D. Gabovich et al., Maximum compression of compensated ion bursts by phase focusing, *Pis'ma Zh. Eksp. Teor. Fiz.*, **6**, 1080–1084 (1980).

37. A. I. Morozov and S.V. Lebedev. Plasma optics, in: *Reviews of Plasma Physics* , Vol. 8, Consultants Bureau, New York (1980), pp. 301–460.

38. R. Booth and M. W. Lefevre, Space charge lens for high current ion beams, *Nucl. Instrum. Methods,* **151**, 143–147 (1978).

39. M. D. Gabovich, I. S. Gasanov, and I. M. Protsenko, Plasma lenses for ion beam formation, Institute of Physics, Ukrainian Academy of Sciences, Kiev, Preprint No. 8 (1982).

40. M. D. Gabovich, *Compensated Ion Beams* [in Russian], Znanie, Moscow (1980), No. 1.

41. G. I. Budker and A. N. Skrinskii, Electron cooling and new prospects in elementary particle physics, *Usp. Fiz. Nauk*, **124,** 561–567 (1978).

42. F. Winterberger, Focusing of intense ion beams by radiation cooling in a magnetic mirror, *Phys. Rev. Lett.*, **37**, 713–717 (1976).

43. N. N. Semashko, Injectors of fast hydrogen atoms, in: *Progress in Science and Engineering, Plasma Physics Series* [in Russian], VINITI, Moscow (1980), Vol. 1, Part 1, pp. 232–282.

44. A. N. Vladimirov et al., Injectors for the INTOR demonstration thermonuclear reactor, in: *Problems in Atomic Science and Engineering, Thermonuclear Fusion Series* [in Russian] (1981), No. 2(8), pp. 27–32.

45. N. N. Semashko et al., The experimental injector IREK, in: *Talks at the Second All-Union Conf. on Problems of Fusion Reactor Injectors*, Leningrad (1981), Vol. 1, pp. 352–354.

46. N. N. Semashko, V. M. Kulygin, and A. A. Panasenkov, Gas discharge emitters of ion sources for fusion devices, *Proc. XV International Conf. on Phenomena in Ionized Gases*, VINITI, Minsk, USSR (1981), pp. 331–340.

47. V. M. Kulygin et al., Ion source without an external magnetic field, *Zh. Tekh. Fiz.*, **49,** 168–172 (1979).

48. A. A. Panasenkov et al., Hydrogen injector with an edge magnetic field, in: *Plasma Accelerators and Ion Injectors* [in Russian], Nauka, Moscow (1984), pp. 154–163.

49. W. L. Stirling, C. C. Tsai, and P. M. Ryan, *Rev. Sci. Instrum.*, **48**, 533–542 (1977).

49a. C. C. Tsai et al., Duopigatron ion source for PLT injectors, in: *Proc. 7th Symp. on Engineering Problems of Fusion Research,* Knoxville, USA (1977), IEEE Publ. No. 77CH1267–4 NPS, pp. 278–283.

50. N. N. Semashko, A. A. Panasenkov, and V. M. Kulygin, Some aspects of the stationary ion sources construction, in: *Proc. 7th Symp. on Engineering*

Problems of Fusion Research, San Francisco, USA (1979), IEEE Publ. No. 79CH1441–5 NPS, pp. 221–224.

51. M. Fumelli and F. P. Valckx, Periplasmatron, an ion source for intensive neutral beams, *Nucl. Instrum. Methods,* **135**, 203–208 (1976).

52. N. V. Pleshivtsev, N. N. Semashko, and I. A. Chukhin, "Antiprobkotron" plasma ion source, Inventor's Certificate No. 294545 (USSR). Published in *Byull. Izobret.,* No. 6 (1973). French Patent No. 71.28566, British Patent 1348562, US Patent 3,798,488.

53. R. Becherer, M. Fumelli, and F. P. Valckx, The rectangular periplasmatron, an ion source for MW neutral beam injection systems, in: *Proc. Seventh Symp. on Engineering Problems of Fusion Research,* Knoxville, USA (1977), IEEE Publ. No. 77CH1267–4 NPS, Vol. 1, pp. 287–290.

54. V. M. Kulygin and A. A. Panasenkov, Particle and energy balance in an ion source discharge, Moscow, Preprint No. IAE–3322 (1980).

55. R. Limpaecher and K. R. MacKenzie, *Rev. Sci. Instrum.,* **44**, 726–742 (1973).

56. K.W. Ehlers, Rectangularly shaped large area plasma source, *Rev. Sci. Instrum.,* **50**, 1353–1362 (1979); **52**, 1473–1435 (1982); *Proc. Seventh Symp. on Engineering Problems of Fusion Research,* Knoxville, USA (1977), IEEE Publ. No. 77CH1267–4 NPS, Vol. 1, pp. 291–294.

57. Y. Okumura, *Rev. Sci. Instrum.,* **55**, 1–8 (1984).

57a. Y. Ohara et al., Ion source development for JT–60 neutral beam injector, in: *Proc. 7th Symp. on Engineering Problems of Fusion Research,* Knoxville, USA (1977), IEEE Publ. No. 77CH1267–4 NPS, Vol. 1, pp. 273–277.

58. V. G. Grigor'yan et al., Fluxes of secondary particles in the ion-optical systems of hydrogen ion sources, in: *Plasma Accelerators and Ion Injectors* [in Russian], Nauka, Moscow (1984), pp. 163–169.

59. *The National Economy of the USSR 1922–1982* [in Russian], Jubilee Statistical Yearbook of the Central Statistical Administration of the USSR, Finansy i Statistika, Moscow (1982).

60. A. L. Pressman, *Corrosion – Enemy and Friend* [in Russian], Znanie, Moscow (1971).

61. M. Kaminsky, *Atomic and Ionic Impact Phenomena on Metal Surfaces,* Springer-Verlag (1965)

62. N. V. Pleshivtsev, *Cathode Sputtering* [in Russian], Atomizdat, Moscow (1968).

63. N. V. Pleshivtsev, *Physical Problems in Cathode Sputtering. A Review* [in Russian], I. V. Kurchatov Institute of Atomic Energy, Moscow (1979).

64. R. Behrisch, ed., *Problems in Applied Physics. Sputtering of Solids by Ion Bombardment. Physical Sputtering of Single-Element Solids* [Russian translation], (1984), 2nd ed. (1986).

65. G. M. McCracken, The behavior of surfaces under ion bombardment, *Rep. Prog. Phys.,* **38**, 241–327 (1975).

66. J. Roth, J. Bohdansky, and W. Ottenberger, Data on low energy light ion sputtering, Max Planck Inst. fuer Plasmaphysik, Garching bei Muenchen, Report No. IPP 9/26 (1979).

67. N. Matsunami et al., Energy dependence of sputtering yields of monatomic solids, Inst. of Plasma Phys., Nagoya Univ., Japan, Report No. IPPJ-AM–14 (1980).

68. J.-C. Piven, Review. An overview of ion sputtering physics and practical applications, *J. Mater. Sci.*, **18**, 1267–1290 (1983).
69. Sh. G. Askerov and L. A. Sena, Cathode sputtering of metals by slow mercury ions, *Fiz. Tverd. Tela*, **11**, 1591–1597 (1969).
70. J. Bohdansky, J. Roth, and H. L. Bay, An analytical formula and important parameters for low-energy ion sputtering, *J. Appl. Phys.*, **51**, 2861–2865 (1980); <u>52</u>, 1610 (1981).
71. J. Roth, J. Bohdansky, and A. P. Martinelli, Low energy light ion sputtering of metals and carbides, *Radiat. Eff.*, **48**, 213–220 (1980).
72. P. Sigmund, Theory of sputtering. 1. Sputtering yield of amorphous and polycrystalline targets, *Phys. Rev.*, **184**, 383–416 (1969).
73. M. Kaminsky, ed., *Radiation Effects on Solid Surfaces, Advances in Chemistry Series 158,* American Chemical Society, Washington, DC (1976).
74. K. Kanaya et al., Consistent theory of sputtering of solid targets by ion bombardment using power potential law, *Jpn. J. Appl. Phys.*, **12**, 1297–1306 (1973).
75. Yu. A. Ryzhov and I. I. Shkarban, Generalization of experimental data on mass exchange between atomic fluxes and polycrystalline surfaces, in: *Topical Collection of Papers from the Sergo Ordzhonikidze Moscow Aviation Institute* [in Russian], Vol. 334 (1975), Moscow, p. 23; in: *Collected Talks given at the 2nd All-Union Symp. on the Interaction of Atomic Particles with Solids*, ONTI IAE, Moscow (1972), pp. 196–199.
76. D.L. Smith, Physical sputtering model for fusion reactor first-wall materials, *J. Nucl. Mater.*, **75**, 20–31 (1978).
77. K. Tsunoyama, T. Suzuki, and Y. Ohashi, Sputtering of iron with ion beams of O_2^+, N_2^+, and Ar^+, *Jpn. J. Appl. Phys.*, **15**, 349–355 (1976).
78. D. L. Smith, *Proc. of Workshop on Sputtering Caused by Plasma Surface Interactions,* Argonne Natl. Lab., Report No. CONF–790775 (1979).
79. V. V. Pletnev, D. S. Semenov, and V. G. Tel'kowsky, On the theory of binary alloy sputtering by light ions, *Radiat. Eff.*, **83**, 113–119 (1984).
80. D. S. Semenov, Method for calculating the sputtering characteristics of first wall materials in thermonuclear plasma experiments [in Russian], Author's Abstract of Candidate's Dissertation in Physical and Mathematical Sciences, MIFI, Moscow (1984).
81. H. H. Andersen and H. L. Bay, Heavy-ion sputtering yields of gold: Further evidence of nonlinear effects, *J. Appl Phys.*, **46**, 2416–2422 (1975).
82. D. A. Thomson, Application of an extended linear cascade model to the sputtering of Ag, Au, and Pt by heavy atomic and molecular ions, *J. Appl. Phys.*, **52**, 982–989 (1981).
83. D. J. Lepoire et al., Sputtering of SO_2 by high energy ions, *Radiat. Eff.*, **71**, 245–259 (1983).
84. C. K. Meins et al., Sputtering of UF_4 by high energy heavy ions, *Radiat. Eff.*, **71**, 13–33 (1983).
85. K. Merkle and W. Jäger, Direct observation of spike effects in heavy-ion sputtering, *Philos. Mag.*, **A44**, 741–762 (1981).
86. D. Pramanik and D. N. Seidman, Atomic resolution observation of nonlinear depleted zones in tungsten irradiated with metallic diatomic

molecular ions, *J. Appl. Phys.*, **54**, 6352–6367 (1983); *Nucl. Instrum. Methods*, **209/210**, 453–459 (1983).

87.   I. A. Baranov, A. S. Krivokhatskii, and V. V. Obnorskii, Mechanism for sputtering of materials by heavy multiply charged ions – fission fragments, *Zh. Tekh. Fiz.*, **51**, 2457–2475 (1981); *At. Energ.*, **54**, 184–188 (1983).

88.   Y. Kitazoe, N. Hiraoka, and Y. Yamamura, Hydrodynamical analysis of non-linear sputtering yields, *Surf. Sci.*, **111**, 381–394 (1981).

89.   Abstracts of talks at the 19-th All-Union Conf. on Emission Electronics, Sections IV and VI, Tashkent, 18-21 Sept. 1984, Izd-vo. TashGU.

90.   O.B. Firsov, Dependence of sputtering of a target on the angle of incidence of the bombarding particles, *Dokl. Akad. Nauk SSSR*, **189**, 302–304 (1969).

91.   J. Bohdansky, et al., Light ion sputtering of fusion reactor materials in dependence of angle of incidence, *J. Nucl. Mater.*, **103/104**, 339–344 (1981).

91a.   J. Bohdansky, et al., Light ion sputtering for H, D, and He in the energy range of 25 keV to 100 keV, *J. Nucl. Mater.*, **111/112**, 717–725 (1982).

92.   Y. Yamamura, Y, Itikawa, and N. Itoh, Angular dependence of sputtering yields of monatomic solids, Institute of Plasma Physics, Nagoya Univ., Japan, Report No. IPPJ-AM-26 (1983).

92a.   Y. Yamamura, N. Matsunami, and N. Itoh, *Radiat. Eff.*, **71**, 65 (1983).

93.   J.S. Colligon, C.M. Hicks, and A.P. Neokleons, Variation of the sputtering yield of gold with ion dose, *Radiat. Eff.*, **18**, 119–126 (1973).

94.   R. Behrisch, et al, Dependence of light-ion sputtering yields of iron on ion fluence and oxygen partial pressure, *J. Nucl. Mater.*, **93/94**, 645–655 (1980).

94a.   J. Bohdansky, Important sputtering yield data for tokamaks: A comparison of measurements and estimates, *J. Nucl. Mater.*, **93/94**, 44–60 (1980).

95.   E.S. Mashkova and V.A. Molchanov, Current trends in ion scattering from solid surfaces, *Radiat. Eff.*, **19**, 29–37 (1973).

96.   E.M. Volin, et al., Effect of ion bombardment on the surface morphology of textured tungsten layers, *Elektron. Tekh., Ser. 6. Mater.*, **9** (158), 6–9 (1981).

97.   T.P. Martynenko, Sputtering of porous materials, *Zh. Tekh. Fiz.*, **38**, 759–760 (1968).

98.   V.V. Morozov, V.S. Kresanov, and V.Ya. Shlyuko, Cathode sputtering of porous lanthanum hexaboride, *Zh. Tekh. Fiz.*, **47**, 2526–2529 (1977).

99.   L.B. Shelyakin, et al., Variation in the sputtering coefficient of ferromagnetic materials near the Curie point, in: *Poverkhn.. Fiz., Khim., Mekh.*, No. 6, pp. 65–69 (1983); Pis'ma *Zh. Eksp. Fiz.*, **21**, 197–199 (1975).

100.   N. V. Drozdova and Yu. V. Martynenko, Surface microstructure during ion bombardment, in: *Problems in Atomic Science and Engineering, Thermonuclear Fusion Series* [in Russian], Izd. IAE im. I. V. Kurchatov, Moscow (1980), Vol. 1(5), pp. 111–113.

101.   R. C. Nelson, An investigation of thermal spikes by studying the high energy sputtering of metals at elevated temperatures, *Philos. Mag.*, **11**, 291–302 (1965).

102.   E. P. Vaulin, N. E. Georgieva, and T. P. Martynenko, Dependence of the sputtering yield of copper on temperature, *Fiz. Tverd. Tela*, **19**, 1423–1425

(1977); *Izv. Leningr. Elektrotekh. Inst.*, RIO LETI, Leningrad, No. 236, 41–42 (1978).

102a. A. P. Koshcheev and A. M. Panesh, Selective sputtering and changes in the electrical and adsorption properties of semiconductor oxide ZnO and TiO_2 surfaces acted on by inert gas ions with energies of 10–300 eV, *Izv. Akad. Nauk SSSR, Ser. Fiz.*, **44**, 428–437 (1980).

102b. J. P. Biersack and E. Santner, Sputtering of potassium chloride by H, He, and Ar ions, *Nucl. Instrum. Methods*, **132**, 229–235 (1976).

102c. G. I. Kostyuk, Critical current densities of charged ions in erosion models, in: *Plasma Sources and Accelerators* [in Russian], Izd-vo Kharkovskogo Aviatsionnogo Inst. im. N. E. Zhukovskogo, Kharkov (1981), No. 5, pp. 87–108.

102d. A. A. Andreev and V. S. Charnysh, Use of ion beams for studying the polymorphic phase transition in cobalt, in: *Ion Beam Diagnostics of Surfaces. Abstracts of Talks* [in Russian], Izd-vo Zaporozhskogo Mashinostr. Inst., Zaporozhe (1983), pp. 7–8.

103. Yu. V. Martynenko, Plasma–Surface Interactions [in Russian], in: V. D. Shafranov, ed., *Progress in Science and Engineering, Plasma Physics Series*, Vol. 3, VINITI, Moscow (1982), pp. 119–175.

104. B. S. Danilin and V. Yu. Kireev, *Ion Etching of Microstructures* [in Russian], Sovetskoe Radio, Moscow (1979).

105. Yu. D. Chistyakov and Yu. P. Rainova, *Physical and Chemical Foundations of Microelectronics Technology* [in Russian], Metallurgiya, Moscow (1979).

106. V. A. Labunov et al., Modern magnetron sputtering devices, *Zarub. Elektron. Tekh.*, No. 10 (256), 3–62 (1982).

106a B. S. Danilin and V. K. Syrchin, *Magnetron Sputtering Systems* [in Russian], Radio i Svyaz', Moscow (1982).

107. G. F. Ivanovskii, The use of dry processes in microelectronics, *Elektron. Promst.*, No. 3, 26–32 (1980).

108. V. P. Belevskii et al., An analysis of electron–ion methods for producing clean metal films, *Prib. Tekh. Eksp.*, No. 5, 182–185 (1973).

109. A. I. Grigorov and A. P. Semenov, *Machining Gas Bearings by Ion Sputtering* [in Russian], Nauka, Moscow (1976).

110. S. Namba, ed., *Ion Doping Technology* [Russian translation from Japanese], V. F. Ovcharova, trans., and P. V. Pavlov, ed., Sovetskoe Radio, Moscow (1974).

111. E. I. Zorin, P. V. Pavlov, and D. I. Tetel'baum, *Ion Doping of Semiconductors* [in Russian], Énergiya, Moscow (1975).

112. V. M. Gusev and M. I. Guseva, Ion doping of semiconductors, *Priroda*, No. 12, 42–52 (1979).

113. V. A. Simonov et al., The "Vezuvii-9" system for high energy ion implantation, *Elektron. Tekh. Ser. 7: Tekhnol., Org. Proizvod., Oborudov.*, No. 1 (110), 71–73 (1982).

114. E. S. Tsyrlin et al., *Metalloved. Term. Obrab. Metal.*, No. 5, 19–23 (1983).

115. Yu. M. Lakhtin and A. D. Kogan, *Nitriding Steels* [in Russian], Mashinostroenie, Moscow (1976).

116. V. S. Barashenkov, *New Tasks for Heavy Ions* [in Russian], Atomizdat, Moscow (1977).

117.   A. A. Babad-Zakhryapin and G. D. Kuznetsov, *Radiation Stimulated Chemical and Thermal Processing* [in Russian], Énergoizdat, Moscow (1982).

117a.   A. A. Babad-Zakhryapin and G. D. Kuznetsov, *Textured High-Temperature Coatings* [in Russian], Atomizdat, Moscow (1980).

117b.   A. A. Babad-Zakhryapin, *High-Temperature Processes in Materials Damaged by Low-Energy Ions* [in Russian], Énergoatomizdat, Moscow (1985).

118.   U. A. Arifov and T. D. Radzhabov, *Sorption Processes during the Interaction of Charged Particles with Solid Surfaces* [in Russian], Izd-vo Fan, Tashkent (1974).

119.   L. Pranyavichyus and Yu. Dudonis, *Modification of the Properties of Solids by Ion Beams* [in Russian], Mokslas, Vilnius (1980).

120.   M. I. Guseva, Ion implantation in metals, *Poverkhnost'*, No. 4, 27–30 (1982).

121.   A. M. Dorodnov, Engineering plasma sources, *Zh. Tekh. Fiz.*, **48**, 1858–1870 (1978).

122.   E. P. Vaulin et al., eds., Abstracts of Talks at the 4th All-Union Conf. on Plasma Sources and Ion Injectors, Vses. Nauchno-Tekh. Tsentr, Moscow (1978).

123.   N. P. Kozlov, ed., Abstracts of Talks at the 5th All-Union Conf. on Plasma Sources and Ion Injectors, Nauka, Moscow (1982).

124.   A. I. Morozov, Yu. M. Levchenko, and A. V. Trofimov, Engineering applications of PIG sources [in Russian], Moscow, Preprint IAE–3282/7 (1980).

125.   N. N. Semashko et al., Possible use of high-current (10–100 A) modular ion sources without external magnetic fields for engineering purposes, Moscow, Preprint IAE–3257/7 (1980).

126.   S. D. Grishin, L. V. Leskov, and N. P. Kozlov, *Plasma Sources* [in Russian], Mashinostroenie, Moscow (1983).

127.   S. D. Grishin, L. V. Leskov, and V. V. Savichev, *Space Technology and Manufacturing* [in Russian], Znanie, Moscow (1978).

128.   A. I. Morozov and A. P. Shubin, *Electrojet Engines for Space* [in Russian], Znanie, Moscow (1975).

129.   V. G. Padalka and V. T. Tolok, Engineering techniques at high energies, *At. Energ.*, **44**, 476–478 (1978).

129a.   M. K. Marakhtanov, *Commercial Ion Systems. A Training Manual* [in Russian], Izd-vo MVTU im. N. E. Baumana, Moscow (1976).

130.   V. M. Golyanov and A. P. Demidov, Inventor's Certificate No. 411037 (USSR), A method for obtaining artificial diamonds, Priority from 28 October 1977. Published in *Byull. Izobret.*, No. 29, 172 (1974). Patents: USA No. 3840451 (1974), UK No. 139987 (1975), Switzerland No. 582622 (1976), France No. 2157957 (1976), W. Germany No. 2252343 (1977), and Japan No. 52–42159 (1977).

131.   V. T. Cherepin, M. A. Vasil'ev, and Yu. N. Ivashchenko, Effect of ion bombardment on the corrosion resistance of iron–carbon alloys, *Dokl. Akad. Nauk SSSR*, **210**, 821 (1973).

132.   Yu. M. Khirnyi and A. P. Solodovnikov, Enhanced corrosion resistance in metals irradiated by helium ions, *Dokl. Akad. Nauk SSSR*, **214**, 82–83 (1974).

133. M. D. Gabovich et al., Hardening of steel by ion bombardment, in: V. T. Cherepin, ed., *The Interaction of Atomic Particles with Solids* [in Russian], Inst. Metallofiziki AN UkrSSR, Naukova Dumka, Kiev (1974), Part 2, pp. 136–137.

134. A. V. Pavlov et al., Changes in the structure and mechanical properties of metals owing to ion bombardment, in: V. T. Cherepin, ed., *The Interaction of Atomic Particles with Solids* [in Russian], Inst. Metallofiziki AN UkrSSR, Naukova Dumka, Kiev (1974), Part 1, pp. 114–117.

135. T. D. Radzhabov, Surface and near-surface sorption phenomena during the interaction of charged particles with solids [in Russian], Author's Abstract of Doctoral Dissertation in Physical and Mathematical Sciences, Moscow (1979).

136. A. P. Vinogradov et al., Detection of the properties of unoxidized ultradisperse forms of simple substances on the surfaces of objects in space, Certificate No. 219, *Otkryt., Izobret., Tovar. Znaki* (1979), pp. 1–2.

137. E.A. Dukhovskoi et al., An anomalously low friction phenomenon, Certificate No. 219, *Otkryt., Izobret., Tovar. Znaki* (1973), No. 13, p. 1; *Dokl. Akad. Nauk SSSR*, **189**, No. 6 (1969).

138. N. N. Petrov and I. A. Abroyan, *Ion-Beam Diagnostics of Surfaces* [in Russian], Izd-vo LGU, Leningrad (1977).

139. Ion-Beam Diagnostics of Surfaces [in Russian], Abstracts of lectures at an All-Union Meeting–Seminar, Uzhgorod, Izd-vo UzGU (1977).

140. V. T. Cherepin and M. A. Vasil'ev, *Methods and Apparatus for Surface Analysis of Materials. A Handbook* [in Russian], Naukova Dumka, Kiev (1982).

141. L. V. Nevzorova, ed., The Interaction of Atomic Particles with Solids. *The Second All-Union Symposium. Collected Talks*, 9–11 October 1972, ONTI IAE (1972).

142. A. G. Koval', ed., The Interaction of Atomic Particles with Solids. *Brief Contents of Talks at the 4th All-Union Conference*, Kharkov State University, Research Laboratory for Ion Processes, 1976, Izd-vo Kharkovskogo Gos-go. Un-ta., Kharkov (1976).

143. V. E. Borisenko, ed., The Interaction of Atomic Particles with Solids. *Materials from the 5th All-Union Conference,* Minsk, 1978, Izd-vo Minskogo Radiotekh. Inst. (1978).

144. V. P. Parkhutik, ed., The Interaction of Atomic Particles with Solids. *Materials from the 6th All-Union Conference,* Minsk, 1981, Izd-vo Minskogo Radiotekh. Inst. (1981).

145. A. K. Polonin, ed., The Interaction of Atomic Particles with Solids. *Materials from the 7th All-Union Conference,* Minsk, 1984, Izd-vo Minskogo Radiotekh. Inst. (1984).

146. *14th All-Union Conf. on Emission Electronics*, Tashkent, 1970. *Abstracts of Talks,* Izd-vo Fan, Tashkent (1970). Section 1, Plenary sessions, pp. 1–50; Section 7, Interaction of Atomic Particles with Solid Surfaces. Symposium on Ion Implantation in Solids, U. A. Arifov, Chairman, pp. 1–120.

147. Yu. G. Ptushinskii, ed., *15th All-Union Conf. on Emission Electronics, Brief Contents of Talks*, Kiev (1973), Inst. Fiziki UkrSSR, ONTI, Kiev (1973).

148.   M. V. Kremkov and I. A. Garaftudinova, eds., *19th All-Union Conf. on Emission Electronics, Abstracts of Talks, Sections IV and VI*, Tashkent (1984), Izd. TashGU, Tashkent (1984).

149.   F. W. Saris and W. F. van der Weg, eds., Atomic Collisions in Solids. *Proc. of the 6th International Conf. on Atomic Collisions in Solids*, Amsterdam, September 22–26, 1975; *Nucl. Instrum. Methods*, **132**, 1–718 (1976).

150.   M. S. Tulisova and G. A. Roganova, eds., *Preliminary Programme and Abstracts of Papers Presented at the 7th International Conf. on Atomic Collisions in Solids*, Moscow, September 19–23, 1977, Moscow State University Publishing House (1977).

151.   G. A. Iferov, ed., *Preliminary Program and Abstracts of Talks at the 10th Conf. on the Application of Charged-Particle Beams for Studying the Composition and Properties of Materials* [in Russian], Moscow, 28–30 May 1979, Izd-vo MGU, Moscow (1979), *12th All-Union Conf.*, Moscow, 31 May–2 June 1982, MGU, Moscow (1982).

152.   W. Schwenke and L. Heunemann, eds., Beitraege zur 7 Tagung Hochvakuum, Grenzflachen/Duenne Schichten, Dresden2–5 Maerz 1981, Physikalische Gesellschaft der DDR (1981).

153.   R. Behrisch et al., eds., *Ion Surface Interaction, Sputtering, and Related Phenomena*, Max Planck Institut fuer Plasmaphysik, Garching bei Muenchen, Germany, 24–27 September, Gordon and Breach Sci. Publ., London–New York–Paris (1972); *Radiat. Eff.*, **18**, No. 1–2 (1973), **19**, No. 1–4 .

154.   S. T. Picraux, E. P. Eernisse, and F. L. Vook, *Applications of Ion Beams to Metals*, Sandia Lab., Albuquerque, New Mexico (1973); Plenum Press, New York–London (1974).

155.   B. I. Crowder, ed., *Ion Implantation in Semiconductors and Other Materials, 3rd International Symposium*, Yorktown Heights, New York, 1972, Plenum Press, New York (1973).

156.   Y. Guylai, T. Lohner, and E. Pasztor, eds., *Proc. 1st Conf. on Ion Beam Modification of Materials*, 4–8 Sept. 1978, Budapest, Hungary (1978), Vol. III, pp. 1465–1508; *Radiat. Eff.*, **48**, No. 1–4, pp. 1–256 (1980).

157.   *Proc. 2nd International Conf. on Ion Beam Modification of Materials*, 14–18 July 1980, Albany, New York; *Nucl. Instrum. Methods*, **182/183**, Part I, 1–544; Part II, 545–1043 (1981).

158.   B. Biasse, G. Destafains, and J. P. Gailliard, eds., *Proc. 3rd International Conf. on Ion Beam Modification of Materials*, 6–10 Sept. 1982, Albany, New York; *Nucl. Instrum. Methods*, **209/210**, 1–1207 (1983).

159.   *Proc. of the 21st National Vacuum Symposium*, 8–11 Oct. 1974, Anaheim, Calif.; *J. Vac. Sci. Technol.*, **12**, 191–351 (1975); **13**, 3–574 (1976).

160.   R. E. Honig, ed., *Proc. of the Symp. on Advances in Ion Technology; Seventh Annual Symp. Sponsored by the Greater New York Chapter of the American Vacuum Society*, 26 May 1976; *J. Vac. Sci. Technol.*, **13**, 1001–1046 (1976); **14**, 17–508 (1977).

161.   *Proc. of 25th Natl. Symp. Am. Vacuum Soc.*, *J. Vac. Sci. Technol.*, **16**, 109–763, 1609–1686 (1979).

162.   *Proc. of 26th Natl. Symp. Am. Vacuum Soc.*, *J. Vac. Sci. Technol.*, **17**, 1–558 (1980).

163. *Proc. of 27th Natl. Symp. Am. Vacuum Soc.*, *J. Vac. Sci. Technol.*, **18**, 149–1400 (1981).
164. *Proc. of 28th Natl. Symp. Am. Vacuum Soc.*, *J. Vac. Sci. Technol.*, **20**, 265–1427 (1982).
165. *Proc. of 29th Natl. Symp. Am. Vacuum Soc.*, *J. Vac. Sci. Technol.*, *Second Series*, **1**, 111–802, 811–1345 (1983).
166. *Proc. of the 13th Annual Symp. of the Greater New York Chapter of the AVS on Plasma and Ion Processing*, *J. Vac. Sci. Technol.*, **21**, 725–903 (1982).
166a. *Proc. of the 1983 International Symp. on Electron, Ion, and Photon Beams*, *J. Vac. Sci. Technol.*, **1B**, 958–1401 (1983).
167. P. D. Townsend, J. E. Kelly, and N. E. W. Hartley, eds., *Ion Implantation, Sputtering and Their Applications*, Academic Press, New York–London–San Francisco (1976).
168. Kh. Rissel and I. Ruge, *Ion Implantation* [Russian translation], Nauka, Moscow (1983).
169. A. P. Dostanko et al., *Plasma Metallization in Vacuum* [in Russian], Nauka i Tekhnika, Minsk (1983).
170. O. Auciello, Ion interaction with solids: Surface texturing, some bulk effects, and their possible applications, *J. Vac. Sci. Technol.*, **19**, 841–867 (1981).
171. H. R. Kaufman, J. J. Cuomo, and J. M. E. Harper, Critical Review. Technology of broadbeam ion sources used in sputtering. Part I. Ion source technology, *J. Vac. Sci. Technol.*, **21**, 725–736 (1982).
171a. J. M. E. Harper, J. J. Cuomo, and H. R. Kaufman, Critical Review. Technology of broadbeam ion sources used in sputtering. Part II. Applications, *J. Vac. Sci. Technol.*, **21**, 764–767, 737–756 (1982); **1A**, 337–339 (1983).
172. H. Liebl, Ion probe microanalysis. Review Article, *J. Phys. E*, **8**, 797–808 (1975); *Anal. Chem.*, **46**, 22A–30A (1974).
173. A. Benninghoven, Surface analysis by means of ion beams. *CRC Critical Reviews in Solid State Science* (1976), June, pp. 291–316.
174. R. E. Honig, Surface and thin film analysis of semiconductor materials, *Thin Solid Films*, **31**, 89–122 (1976).
175. A. I. Morozov et al., Application of ion accelerators for plasma cleaning of steel strip surfaces, in: *4th All-Union Conf. on Plasma Accelerators and Ion Injectors. Abstracts of Talks*, Vses. Nauchno-tekhn. Informats. Tsentr., Moscow (1978), pp. 413–414.
176. V. M. Artemov, D. V. Iremashvili, and A. I. Shimko, Gun for cleaning solid surfaces by ion bombardment, *Prib. Tekh. Eksp.*, No. 1, 237–238 (1973).
177. S. Schiller, U. Heisig, and K. Steinfelder, A new sputter cleaning system for metallic substrates, *Thin Solid Films*, **33**, 331–339 (1976).
178. G. L. Jackson et al., Initial wall conditioning in Doublet III, *J. Vac. Sci. Technol.*, **1A**, 1861–1867 (1983).
179. G. Paszti et al., Preliminary results of the investigation of plasma contamination in MT–1 tokamak on probes by RBS and channeling, Central Research Inst. for Physics, Hungarian Academy of Sciences, Budapest, Report KFKI–1981–27, p. 23.

180. J. E. Houston and R. D. Bland, Relationship between sputter cleaning parameters and surface contaminants, *J. Appl. Phys.*, **44**, 2504–2508 (1973).

181. R. Johnson, G. L. Thomas, and T. Kent, The etching of fatty acid films in a glow discharge, *Thin Solid Films*, **38**, L13-L15 (1976).

182. H. P. Smith, D. W. DeMichele, and J. M. Kahn, Effect of thin carbonaceous films on 500-keV helium ion sputtering of copper, *J. Appl. Phys.*, **36**, 1952–1957 (1965).

182a. N. V. Pleshivtsev, Sputtering of copper by hydrogen ions with energies of up to 50 keV, *Zh. Eksp. Teor. Fiz.*, **37**, 1233–1240 (1959).

183. E. Taglauer, U. Beitat, and W. Heiland, Investigation of ion impact desorption of atoms and molecules by low energy ion scattering (ISS), *Nucl. Instrum. Methods*, **149**, 605–608 (1978).

184. T. Smith, Sputter cleaning and etching of crystal surfaces (Ti, W, Si) monitored by Auger spectroscopy, ellipsometry, and work function change, *Surf. Sci.*, **27**, 45–59 (1971).

185. W. O. Hofer and P. J. Martin, On the influence of reactive gases on sputtering and secondary ion emission. Oxidation of titanium and vanadium during energetic particle irradiation, *Appl. Phys.*, **16**, 271–278 (1978).

186. J. D. Chinn, I. Adesida, and E. D. Wolf, Chemically assisted ion beam etching for submicron structures, *J. Vac. Sci. Technol.*, **1B**, 1028–1032 (1983).

187. L. V. Vishnevskaya et al., Interaction of ions of fluorine-containing gases with the surface of optical glasses, *Opt.-Mekh. Promst.*, No. 7, 30–32 (1981).

188. G. A. Lincoln et al., Large area ion beam assisted etching of GaAs with high etch rates and controlled anisotropy, *J. Vac. Sci. Technol.*, **1B**, 1043–1046 (1983).

189. D. C. Flanders and N. N. Efremov, Generation of <50 nm period gratings using edge defined techniques, *J. Vac. Sci. Technol.*, **1B**, 1105–1108 (1983).

190. B. Holliday, World-Wide Directory of Manufacturers of Vacuum Plant, Components, and Associated Equipment – 1983, *Vacuum*, **33**, 455–502 (1983).

191. Ion beam thinning attachment, type IBT 200, Edwards High Vacuum, England.

192. H. Itakura, H. Komiya, and K. Ukai, Multi-chamber dry etching system, *Solid State Technol.*, 209–214, April 1982.

193. D. L. Flamm et al., XeF_2 and F atom reactions with Si: Their significance for plasma etching, *Solid State Technol.*, 117–121, April 1983.

194. D. T. Hawkins, Ion milling (ion-beam etching), 1954–1975: A bibliography, *J. Vac. Sci. Technol.*, **12**, 1389–1398 (1975); **16**, 1051–1071 (1979).

195. P. H. Schmidt, E. G. Spencer, and E. M. Walters, Ion milling of magnetic oxide platelets for the removal of surface and near-surface imperfections and defects, *J. Appl. Phys.*, **41**, 4740–4742 (1970).

196. L. F. Johnson, K. A. Ingersoll, and D. Kahng, Planarization of patterned surfaces by ion beam erosion, *Appl. Phys. Lett.*, **40**, 636–638 (1982).

197. W. A. Johnson, J. C. North, and R. Wolfe, Differential etching of ion-implanted garnet, *J. Appl. Phys.*, **44**, 4753–4757 (1973).

198. R. K. Watts et al., Microfabrication of circuits for magnetic bubbles of diameter 1 μm and 2 μm, *Appl. Phys. Lett.*, **28,** 355–358 (1976).

199. H. L. Garvin et al., Ion beam micromachining of integrated optics components, *Appl. Opt.*, **12,** 455–459 (1973).

200. R. Adde et al., Ion beam machining of niobium weakly superconducting microbridges, *Rev. Phys. Appl.*, **9,** 179–181 (1974).

201. A. P. Janssen and J. P. Jones, The sharpening of field emitter tips by ion sputtering, *J. Phys. D*, **4,** 118–123 (1971).

202. H. Yasuda, Figuration of wedge-shaped and parabolic surfaces by ion etching, *Jpn. J. Appl. Phys.*, **12,** 1139–1142 (1973).

203. J. B. Schroeder, H. D. Dieselman, and J. W. Douglass, Technical feasibility of figuring optical surfaces by ion polishing, *Appl. Opt.*, **10,** 295–300 (1971).

204. I. G. Bunin, G. B. Rozhnov, and G. S. Khodakov, Shaping of surfaces with a small-area ion beam, *Opt.-Mekh. Promst.*, No. 2, 36–39 (1983).

205. S. Ya. Lebedev and S. D. Panin, Effect of temperature on porosity in nickel during irradiation by nickel ions, *Atom. Energ.*, **41,** 428–429 (1976).

206. V. N. Bogomolov et al., Creation of artificial channel dielectric matrices and production of ultrathin metal grids, *Pis'ma Zh. Tekh. Fiz.*, **8,** 1393–1395 (1982).

207. L. B. Shelyakin et al., Revelation of stressed states in metals by ion bombardment, *Poverkhnost'*, No. 4, 51–61 (1982).

208. D. A. Kiewit, Microtool fabrication by etch pit replication, *Rev. Sci. Instrum.*, **44,** 1741–1742 (1973).

209. G. P. Airey and G. P. Sabol, Transmission electron microscopy of ion-thinned oxides forming during corrosion of Zircalloy–4, *J. Nucl. Mater.*, **45,** 60–62 (1972/73).

210. D. Quataert and C. A. Busse, Investigation of the corrosion mechanism in tantalum–lithium high temperature heat pipes by ion analysis, *J. Nucl. Mater.*, **46,** 329–340 (1973).

211. I. Kh. Chepovetskii, *Elements of Finishing Processing* [in Russian], Naukova Dumka, Kiev (1980).

212. G. A. Muranova et al., Ion polishing of optical coatings, *Opt.-Mekh. Promst.*, No. 5, 33–35 (1979).

213. V. F. Zhiglo and B. A. Terekhov, Polishing of copper by argon ions, in: *Problems of Atomic Science and Engineering. Series on Physical Experimental Technique* [in Russian], No. 2, pp. 40–43 (1980).

214. N. V. Pleshivtsev, G. D. Kuznetsov, and M. L. Margolis, Ion-beam finishing and polishing of metal surfaces, in: *Abstracts of Talks at the Seventh All-Union Conf. on the Interaction of Atomic Particles with Solids*, Minsk (1984), pp. 39–40.

215. A. N. Luzin and A. N. Tyushev, Sputtering of crystals and surface migration of atoms during bombardment by slow ions, *Zh. Tekh. Fiz.*, **49,** 2671–2679 (1979); in: V. P. Parkhutik, ed., The Interaction of Atomic Particles with Solids. *Materials from the 6th All-Union Conference*, Minsk, 1981, Izd-vo Minskogo Radiotekhn. Inst. (1981), Part I, pp. 32–34.

216. D. W. Tomkins and D. S. Coleman, Ion-bombardment etching techniques as applied to powder metallurgy microstructures, *Powder Metall.*, **18,** 283–302 (1975).

216a. D. M. Skorov et al., Cathode-vacuum etching of uranium in the VUP–2K/ system, *At. Energ.*, **32**, 319–320 (1972).

217. L. B. Shelyakin, R.-D. G. Schultze, and V. E. Yurasova, Some features of the formation of surface structure on monocrystals during ion bombardment, *Fiz. Plazmy*, **1**, 488–495 (1975).

218. A. G. Belikov et al., Changes in the surface morphology of Ni, V, and Nb during irradiation by a helium plasma, in: *Problems of Atomic Science and Engineering. Series on Radiation Damage and Radiation Materials Science* [in Russian], No. 2(25), pp. 57–60 (1983).

219. F. Vasiliu, Sputtering erosion, apparent growth and equilibrium of triangular microprofiles submitted to ion bombardment at oblique incidence, *Rev. Roum. Phys.*, **22**, 523–536 (1977).

219a. F. Vasiliu, Oblique incidence sputtering of triangular microprofiles, *Radiat. Eff.*, **45**, 213–218 (1980).

220. C. J. Pellerin et al., Quantitative surface profilometry applied to sputter ion bombarded sapphire, *J. Vac. Sci. Technol.*, **12**, 496–500 (1975).

221. E. G. Spencer and P. H. Schmidt, Ion machining of diamond, *J. Appl. Phys.*, **43**, 2956–2958 (1972).

222. K. M. Polivanov and A. L. Frumkin, Thin magnetic films, in: *Great Soviet Encyclopedia* [in Russian], 3rd ed., Sovet-skaya Éntsiklopediya (1974), Vol. 15, pp. 174–175.

223. J. R. Cavaler, Properties of sputtered high T_c thin films, *J. Vac. Sci. Technol.*, **12**, 103–106 (1975).

224. A. A. Vasenkov and I. E. Efimov, Microelectronics, in: *Great Soviet Encyclopedia* [in Russian], 3rd ed., Sovet-skaya Éntsiklopediya (1974), Vol. 16, pp. 246–248.

225. F. L. Carter, Molecular level fabrication techniques and molecular electronic devices, *J. Vac. Sci. Technol.*, **1B**, 959–968 (1983).

226. D. M. Mattox, Thin film metallization of oxides in microelectronics, *Thin Solid Films*, **18**, 173–186 (1973).

227. V. P. Belevskii, A. I. Kuzmichev, and V. I. Mel'nik, Electron–ion devices for deposition of thin film coatings, *Obshchestvo "Znanie" UkrSSR,* Kiev (1982).

228. A. V. Balakov and A. V. Konshina, Methods for obtaining and the properties of carbon diamondlike films. A review, *Opt.-Mekh. Promst.*, No. 9, 52–59 (1982).

229. V. M. Golyanov and A. P. Demidov, A device for ion sputtering, Inventor's certificate No. 603701 (USSR), Priority from 26 December 1972. Published in *Byull. Izobret.*, No. 15, 86–87 (1978). Patents USA No. 4,049,533 (1977), UK No. 1,480,564 (1977), Switzerland No. 594,742 (1978), and France No. 2,323,774 (1978).

230. V/O Litsenzintorg, Artificial diamond. A method for producing it [in Russian].

231. A. A. Teplov et al., The superconducting transition temperature, critical magnetic fields, and structure of vanadium films, *Zh. Eksp. Teor. Fiz.*, **71**, 1122–1128 (1976).

232. B. V. Deryagin et al., Growth of polycrystalline diamond films from the gaseous phase, *Zh. Eksp. Teor. Fiz.*, **69**, 1250–1252 (1975).

233. A. S. Bakai and V. E. Strel'nitskii, The structure of carbon films formed by deposition of fast ions, *Zh. Tekh. Fiz.*, **51**, 2414–2416 (1981).

234. D. Aisenberg and R. Chabot, Ion-beam deposition of thin films of diamondlike carbon, *J. Appl. Phys.*, **42**, 2953–2961 (1971).

235. B. A. Banks and S. K. Rutledge, Ion beam sputter-deposited diamondlike films, *J. Vac. Sci. Technol.*, **21**, 807–814 (1982).

236. R. S. Nelson et al., Diamond synthesis internal growth during C^+ ion implantation, *Proc. R. Soc. London*, **A386**, 211–222 (1983).

237. T. E. Derry and J. P. E. Sellschop, Ion implantation of carbon in diamond, *Nucl. Instrum. Methods*, **191**, 23–26 (1981).

238. S. Aisenberg and R. W. Chabot, Physics of ion plating and ion beam deposition, *J. Vac. Sci. Technol.*, **10**, 104–107 (1973).

239. Leybold-Heraeus, GMBH, W. Germany, Kathoden-Zerstaubung. Cathode sputtering, 10 November 1981.

239a. Leybold-Heraeus, GMBH, W. Germany, Apparatus, components, materials, and systems for thin film technology, 5 August 1983.

240. Balzers AG, Liechtenstein, Planar magnetron sputtering sources.

240a. Balzers AG, Liechtenstein, LLS801 Load lock sputtering system for integrated circuit metallization.

241. NEVA Corp., Automatic system for sputtering, Model No. ASP–600, Rikkei Corp, Japan.

241a. NEVA Corp., System and continuous production line for sputtering, Model No. ILS–700, Rikkei Corp, Japan.

242. VEB Elektromat, Dresden, Rf system for cathode sputtering, Model HZM–4, January 1973.

243. G. A. Kowalsky, J. P. Maishev, and J. A. Dmitriev, Ion source, U.S. Patent No. 4,122,347, Oct. 24, 1978; Int. Cl² HO1 J 27/100, U.S.Cl. 250/423R.

243a. G. A. Kovalsky et al., Apparatus for ion plasma coating of articles, U.S. Pat No. 4,145,029, Nov. 20, 1979, Int. Cl. C23C 15/00, U.S.Cl. 204/298. British Pat. No. 1,544,612, Filed 4 Jan. 1978, Publ. 25 Apr. 1979.

243b. Yu. P. Maishev, Intense ion beam sources with compensation of positive space charge inside the accelerating gap, *Prib. Tekh. Eksp.*, No. 1, 183–186 (1980).

244. I. G. Kesaev and V. V. Pashkova, Electromagnetic locating of the cathode spot, *Zh. Tekh. Fiz.*, **29**, 287–298 (1959).

244a. V. E. Minaichev et al., Electrical discharge device for thin film deposition, Inventor's certificate No. 426540 (USSR), Author's disclosure No. 1699206 with priority from 20 Sept. 1971; published in *Byull. Izobret.*, No. 37 (1975).

245. J. W. Patten et al., Recent advances in materials synthesis by sputter deposition, *Thin Solid Films*, **83**, 3–4 (1981).

246. O. F. Poroshin and Zh. Zh. Kutan, *Production and Study of Intense Hydrogen Ion Beams* [in Russian], Izd. FTI GK IAE SSSR, Sukhumi (1963).

246a. M. von Ardenne, Tabellen zur Angewandten Physik, Bd. 2, VEB Deutscher der Wissenschaften, Berlin (1962), pp. 640–705.

246b. S. N. Popov, A study of the Ardenne duoplasmatron, *Prib. Tekh. Eksp.*, No. 4, 20–24 (1961).

246c. R. A. Demirkhanov, Yu. V. Kursanov, and V. M. Blagoveshchenskii, High intensity proton source, *Prib. Tekh. Eksp.*, No. 2, 19–21 (1964).

247.   N. V. Pleshivtsev et al., Helium ion beam with a current of 200 mA and an energy of 70 keV, *At. Energ.*, **22**, 128–131 (1967).

248.   W. A. S. Lamb and E. J. Lofgren, High-current ion injector, *Rev. Sci. Instrum.*, **27**, 907–909 (1956).

249.   N. V. Pleshivtsev et al., Cw injector of 0.5-A, 115-keV hydrogen ions and 0.15-A, 75-keV helium ions, *Prib. Tekh. Eksp.*, No. 6, 23–28 (1967).

249a. H. Tawara, Suganomata, and S. Suematsu, A compact duoplasmatron ion source using a ferrite permanent magnet, *Nucl. Instrum. Methods*, **31**, 353–354 (1964).

250.   N. V. Pleshivtsev et al., The production of intense ion beams using a magnetized plasma, *Plasma Phys.*, **10**, 45–53 (1968).

250a. N. V. Pleshivtsev and B. M. Bezverkhov, Graphical-analytical calculation of an ion beam taking space charge into account, *Radiotekh. Elektron.*, No. 5, 829–838 (1969).

251.   V. N. Danilov, Bipolar current in a special magnetic field, in: *Studies on Physics and Electronics, Proceedings of the Moscow Physicotechnical Institute* [in Russian], Oborongiz, Moscow (1962), No. 10, pp. 67–79.

252.   I. A. Abroyan, A. N. Andronov, and A. I. Titov, *The Physical Foundations of Electron and Ion Technology* [in Russian], Vysshaya Shkola, Moscow (1984).

253.   P. H. Schmidt, R. N. Castellano, and E. G. Spencer, Deposition and evaluation of thin films by DC ion beam sputtering, *Solid State Technol.*, **15**, 27–31 (1972).

254.   S. M. Kane and K. Y. Ahn, Characteristics of ion-beam-sputtered thin films, *J. Vac. Sci. Technol.*, **16**, 171–174 (1979).

255.   T. C. Tisone and P. D.Cruzan, Low-voltage triode sputtering with a confined plasma: Part V – Application to backsputter definition, *J. Vac. Sci. Technol.*, **12**, 677–688 (1975).

256.   J. L. Vossen, A sputtering technique for coating the inside walls of through holes in substrates, *J. Vac. Sci. Technol.,* **11**, 875–877 (1974).

257.   P. Haymann, P. Rodocanachi, and M. Meyer, Condensation d'faisceau atomique d'or sur une surface monocristalline, *Entropie*, No. 18, 68–72 (1967).

258.   G. Sletten and P. Knudsen, Preparation of isotope targets by heavy ion sputtering, *Nucl. Instrum. Methods*, **102**, 459–463 (1972).

259.   W. D. Westwood and S. J. Ingrey, Fabrication of optical waveguides by ion-beam sputtering, *J. Vac. Sci. Technol.*, **13**, 104–106 (1967).

260.   M. Garrigues et al., Metal/insulator/Si structures with low interface state density fabricated by combined ion beam sputtering and atomic hydrogen beam treatment, *J. Appl. Phys.*, **54**, 2863–2865 (1983).

261.   R. D. Ivanov, *Magnetic Metal Films in Microelectronics* [in Russian], Sovetskoe Radio, Moscow (1980).

262.   H. Gill and M. P. Rosenblum, Magnetic and structural characteristics of ion beam sputtering deposited Co–Cr thin films, *IEEE Trans. Magn.*, **MAG–19**, 1644–1646 (1983).

263.   G. I. Rukman, Ya. A. Yukhvidin, and I. A. Kalyabina, A method for obtaining thin films, Authors' certificate No. 109057, 48v, 17^{00}, S 23/s (USSR), Priority 28 October 1955. Published in *Byull. Izobret.*, No. 10 (1957).

264.   J. Amano and R. P. W. Lawson, Thin-film deposition using low-energy ion beams (2). Pb$^+$ ion-beam deposition and analysis of deposits, *J. Vac. Sci. Technol.*, **14**, 690–694 (1977).

265.   D. M. Mattox, Fundamentals of ion plating, *J. Vac. Sci. Technol.*, **10**, 47–52 (1973).

266.   S. Schiller, U. Heisig, and G. Goedicke, Alternating ion plating. A method of high-rate ion vapor deposition, *J. Vac. Sci. Technol.*, **12**, 858–864 (1975).

267.   T. D. Radzhabov et al., Interaction of atoms with defects and the formation of heterostructures during ion implantation of impurities in growing films and layers, *Zh. Tekh. Fiz.*, **51**, 1219–1228 (1981); in: M. V. Kremkov and I. A. Garaftudinova, eds., *XIX All-Union Conf. on Emission Electronics, Abstracts of Talks, Sections IV and VI,* Tashkent (1984), Izd. TashGU, Tashkent (1984), p. 109.

267a.   T. D. Radzhabov et al., Capture of impurity atoms by defects and the distribution of complexes during ion bombardment of growing films, *Zh. Tekh. Fiz.*, **52**, 2238–2244 (1982).

267b.   A. A. Iskanderova, A. I. Kamardin, and T. D. Radzhabov, Changes in the physical and mechanical parameters of metal films during ion bombardment, in: M. V. Kremkov and I. A. Garaftudinova, eds., *XIX All-Union Conf. on Emission Electronics, Abstracts of Talks, Sections IV and VI,* Tashkent 18–21 Sept.1984, Izd. TashGU, Tashkent (1984), p. 109.

268.   A. S. Derevyanko et al., Structure of formations appearing during simultaneous incidence on a monocrystalline NaCl surface of a flux of hydrocarbon molecules and a beam of Ne$^+$ ions, *Kristallografiya*, **20**, 803–805 (1975).

269.   L. Pranyavichus, Yu. Dudonis, and V. Titas, Effect of an electric field on the kinetics of formation of dielectric films during ion implantation, *Litov. Fiz. Sb.*, **16**, 609–615 (1976).

270.   Yu. V. Bykov and M. B. Guseva, Epitaxial growth of antimony films during ion irradiation, *Pis'ma Zh. Tekh. Fiz.*, **1**, 485–487 (1975).

271.   V. O. Babaev, Yu. V. Bykov, and M. V. Gusev, Effect of ion irradiation on the formation structure and properties of thin metal films, *Thin Solid Films*, **38**, 1–8 (1976).

272.   R. F. Jannick, C. R. Heiden, and A. E. Guttensohn, A rapid loading and unloading coliseum type fixture for ion plating reedblade contacts, *J. Vac. Sci. Technol.*, **11**, 535–536 (1974).

273.   Yu. T. Miroshnicheniko et al., Study of the properties of molybdenum condensates obtained by vacuum-plasma deposition. Part 1. The effect of the ion energy on the microstructure and level of microstresses. in: *Problems in Atomic Science and Engineering. Series on Radiation Damage and Radiation Materials Science* [in Russian], KhFTI AN UkrSSR No. 2(25), pp. 83–91 (1983).

274.   M. A. Napadov, A. L. Sapozhnikov, and M. M. Maslennikov, Prophylaxis of complications in the application of prostheses made of base metals, in: *Abstracts of the 6th Congress of Stomatologists of the Ukrainian SSR,* Poltava, 1984. Zdorov'e, Kiev (1984), p. 186; *Pravda*, 2 July 1984.

275.   A. I. Grigorov et al., PUSK–77–1 system for deposition of ion–vacuum wear-resistant coatings, *Tekhnol. Avtomob.*, No. 6, 10–11 (1978); No. 12, 10–15.

276.   R. Connoly, Reagan's Military Budget [Russian translation], Élektronika, No. 3, 79–86 (1983); The market for the electronics industry in the USA, Western Europe, and Japan in 1976 and 1984. Technical and economic reviews, *Elektronika,* No. 1, 23–60 (1976); No. 1, 27–67 (1984).

277.   N. V. Pleshivtsev, N. P. Malakhov, and N. N. Semashko, Problems in the use of refractory and rare metals in megawatt ion and atom beam injectors for thermonuclear experiments and reactors, in: E.M. Savitskii, M. A. Tylkina, and N.K. Kazanskii, *Research on and Applications of Refractory Metal Alloys (Rhenium, Tungsten, Molybdenum)* [in Russian], Nauka, Moscow (1983), pp. 37–42.

278.   K. Yano et al., Ionization of nitrogen cluster beam, *Jpn. J. Appl. Phys.,* **14**, 526–532 (1975).

279.   T. Takagi, I. Yamada, and A. Sasaki, Ionized-cluster beam deposition, *J. Vac. Sci. Technol.,* **12**, 1128–1134 (1975).

280.   K. Morimoto, Y. Utamura, and T. Takagi, Ionized-cluster beam deposition process for fabricating p–n junction semiconductor layers, U.S. Pat No. 4,161,418, July 17, 1979. U.S. Cl. 148/175, Int Cl ^{2}H01 121/203.

281.   H. Dibbert and G. Meyer-Kretschmer, Laser enrichment of uranium, *At. Tekh. Rubezh.,* No. 2, 23–26 (1976).

282.   N. V. Karlov et al., Atomic vapor sources for laser isotope separation, *Tr. Fiz. Inst., Akad. Nauk SSSR,* **114**, 24–37 (1979).

283.   B. E. Paton et al., Properties and principles of ion-beam welding, *Avtom. Svarka,* **26**, No. 10, 1–4 (1973).

284.   B. E. Paton et al., On deep alloying during ion beam welding, *Dokl. Akad. Nauk SSSR,* **239**, 576–578 (1978).

285.   M. D. Gabovich, V. Ya. Poritskii, and I. M. Protsenko, Cathode sputtering and "ion discharge" in the layer between a highly ionized plasma and an adjacent collector, *Ukr. Fiz. Zh.,* **26**, 164–165 (1981).

286.   B. E. Paton et al., Welding with alloying and ultrasonic processing by an ion beam, *Dokl. Akad. Nauk SSSR,* **273**, 104–106 (1984).

287.   M. D. Gabovich and V. Ya. Poritskii, Nonlinear waves on a liquid metal surface in an electric field, *Pis'ma Zh. Eksp. Teor. Fiz.,* **33**, 320–324 (1981).

287a. M. D. Gabovich and V. Ya. Poritskii, Mechanism for excitation of nonlinear capillary waves on the surface of a liquid metal in contact with a dense plasma, *Zh. Eksp. Teor. Fiz.,* **85**, 146–159 (1983).

288.   M. D. Gabovich et al., Ion-beam welding of VT–6 titanium alloy, *Avtom. Svarka,* **33**, 69–70 (1980).

289.   C. A. Anderson and I. H. Hinthorne, Ion microprobe mass analyzer, *Science,* **175**, 853–860 (1972).

290.   M. D. Gabovich, I. S. Gasanov, and I. M. Protsenko, Extraction of ions from a duoplasmatron through a channel with a radius on the order of the Debye length, *Pis'ma Zh. Tekh. Fiz.,* **24**, 1509–1512 (1980).

291.   A. E. Banner and B. P. Stimpson, A combined ion probe spark source analysis system, *Vacuum,* **24**, 511–517 (1974).

292.   A. R. Hill, Uses of fine focused ion beams with high current density, *Nature,* **218**, 202–203 (1968).

293.   A. Wagner and T. M. Hall, Liquid gold ion source, *J. Vac. Sci. Technol.,* **16**, 1871–1874 (1979).

294. P. Sudraud, C. Colliex, and J. Walle, Energy distribution of EHD emitted gold ions, *J. Phys. Lett*, **40**, No. 9, L–207 (1979).

295. P. D. Prewett, D. K. Jeffries, and I. D. Cockhill, Liquid metal source of gold ions, *Rev. Sci. Instrum.*, **52**, 562–566 (1981).

296. G. Taylor, Disintegration of water drops in an electric field, *Proc. R. Soc. A*, **280**, 383–397 (1964).

297. L. D. Landau and E. M. Lifshits, *Electrodynamics of Continuous Media* [in Russian], GITTL, Moscow (1957).

298. V. E. Krohn and G. R. Ringo, Ion source of high brightness using liquid metal, *Appl. Phys. Lett.*, **27**, 479–481 (1975).

299. I. Yu. Bartashyus, L. I. Penevichyus, and G. N. Fursei, Explosive emission of a liquid gallium cathode, *Zh. Tekh. Fiz.*, **41**, 1943–1948 (1971).

300. M. D. Gabovich, Liquid-metal ion emitters, *Usp. Fiz. Nauk*, **140**, 136–150 (1983).

301. R. Gomer, On the mechanism of liquid metal electron and ion sources, *Appl. Phys.*, **19**, 365–375 (1979).

302. A. Wagner, The hydrodynamics of liquid metal ion sources, *Appl. Phys. Lett.*, **40**, 440–442 (1982).

303. M. D. Gabovich and V. N. Starkov, The field in a spherical diode and the current from a liquid-metal ion emitter, *Zh. Tekh. Fiz.*, **52**, 1249–1251 (1982).

304. R. L. Seeliger et al., A high-intensity scanning ion probe with submicrometer spot size, *Appl. Phys. Lett.*, **34**, 310–312 (1979).

305. A. Wagner, Applications of focused ion beams to microlithography, *Solid State Technol.*, 97–103, May 1983.

306. K. A. Valiev et al., Physical processes of defect formation and etching during ion lithography of silicon dioxide films, *Mikroelektronika*, **11**, 323–328 (1982).

307. K. A. Valiev et al., Etching of positive electron x-ray resists during irradiation by ions at medium energies, *Mikroelektronika*, **12**, 195–199 (1983).

308. R. Clampitt, K. L. Aitken, and D. K. Jefferies, Intense field emission ion source of liquid metal, *J. Vac. Sci. Technol.*, **12**, 1208 (1975).

309. B. I. Ivanov, A possible way of creating high-current ion beams, *Pis'ma Zh. Eksp. Teor. Fiz.*, **20**, 170–173 (1974).

310. B. I. Ivanov et al., A study of autoresonant ion acceleration on experimental models and the development of a field emission injector, Kharkov Physicotechnical Institute, Preprint No. 80–8 (1980).